Shrinking Giants: How Tiny Materials are Revolutionizing Computing

Nancy

TABLE OF CONTENTS

Chapter 1 : Introduction

1.1 Background and Significance

Nowadays, the rapidly growing demand for computing power is driven by the increasingly connected devices that generate massive data. Around fifty years ago, Moore's law predicted that the number of transistors in the electronic circuits would double every two years. Silicon-based transistors have shrunk smaller and smaller to fit this demand trend. However, when the silicon transistors are scaling down to a few nanometers, they are reaching their physical limit, the short channel limits the electron travel, and the surface defect considerably slows the charge flow[10]. Thus, the evolution in transistor structures and adopting new materials is critical to realize the next generation of electronic devices.

Two-dimensional (2D) materials with the layered structure are an excellent candidate to further shrunk the transistor to the atomic scale with their monolayer thickness. The lack of one dimension helps avoid the short channel effect, while the charges are restricted in two dimensions. The extremely flat surface of 2D materials minifies the scattering of charge leading to a high mobility. For example, single-layer graphene has a high intrinsic carrier mobility that enables its applications in various areas[11], however, the lack of natural energy gap makes it not suitable for the 2D semiconductor application. Alternatively, transition metal dichalcogenides (TMDCs), like molybdenum disulfide (MoS_2) and tungsten diselenium (WSe_2), which have a direct band gap at the monolayer structure, has been demonstrated excellent performance in optical and electrical devices[12-14].

However, to apply 2D TMDCs in the practical application, more works need to be done.

Li and his colleagues employed in the Taiwan Semiconductor Manufacturing Company have proposed that the growth of large-scale and high-quality 2D TMDCs is one of the challenges for the industrial 2D transistor.[10] Currently, there are two approaches to fabricate the few-layer 2D TMDCs; one is to peel the monolayers from bulk crystal chemically or mechanically, which is also known as the "top-down" method. This method provided high-quality 2D materials for many fundamental researches investigating novel physical properties. However, the exfoliated materials are random in size, orientations, and thickness, which is inapposite for industrial application. Another approach is the 'bottom-up" method, in which vapor-phase materials deposit on target substrates. For example, the chemical vapor deposition (CVD) method has been demonstrated to grow various 2D TMDCs, such as MoS_2 and WSe_2 flakes on sapphire substrates[6, 7]. By applying metal-organic precursors in the CVD process, wafer-scale and homogeneous 2D TMDCs films exhibit remarkable electronic properties[15, 16]. Moreover, the heterostructures and superlattices of 2D TMDCs have been successfully achieved by a modified CVD approach[17]. In principle, the CVD method has the potential to produce wafer-scale and high-quality 2D TMDCs at a low cost. Whereas, the CVD process is complicated with many parameters, such as the growth temperature, precursor ratio, and gas environment, which makes the growth mechanism still unclear. Thus, it needs more exploration to the materials research to realize the practical application.

1.2 Present Challenges

To date, significant effort has been reported to grow wafer-scale and high-quality 2D TMDCs through the CVD method. Nevertheless, there still remain several challenges that need to be overcome. First, it is hard to control the layer number through the CVD approach, which is critical to the electronic properties of 2D TMDCs. Although researchers have proposed kinetic and thermodynamic control of layer number of TMD flakes[18, 19], the large-scale layer-controllable TMD film still not accessible. Second, the stacking of hetero-structure vertically and laterally of 2D TMDCs with abrupt interface still out of control. Last but not least, the quality of CVD-grown TMDCs is not good enough to show its intrinsic properties. The defect density and grain boundary in the 2D TMDCs slow the charge flow, affecting the electronic device performance. Therefore, the mechanism study and modified approach are critical to realizing the scalability, crystallinity, and controllability of CVD-grown 2D TMDCs.

1.3 Objective and Contributions

First, we intend to study the growth mechanism of 2D TMDCs in the CVD process to realize layer controllability. In Chapter 4, a two-step CVD approach was systematically applied to investigate the growth mode control under high tensile strain. We found that the precursor ratio was strongly associated with the strain level on the first epilayer, which tunes the growth mode of overlayer TMDCs as Layer-by-Layer Growth. Follow this growth mechanism, WSe_2 bilayers and trilayers were successfully achieved with 100% yield. We further applied this mechanism to the heterostructure system and successfully obtained MoS_2/WSe_2 vertical heterostructures. Our results proved that the tensile strain could direct the TMDCs growth mode from layer-plus-island to layer-by-layer growth, which is critical for large area layer number constructability.

To realize the 2D semiconductor industry, we plan to investigate a new CVD system for high-quality TMDCs materials growth. In chapter 5, we designed a chemical vapor transport (CVT) setup with high-purity W source to achieve high-mobility WS_2 monolayers. As-grown monolayer WS_2 exhibited excellent optical properties with high intensity and narrow full width half maximum. The field-effect transistor performance showed high carrier mobility (the highest as ~200 cm^2 $V^{-1}s^{-1}$) and high current density (520 µA/µm) at room temperature. Through studying the STM results, we found that the defect density of our CVD process is comparable to the exfoliated samples.

Chapter 2 : Literature Review

2.1 Introduction to 2D Transition Metal Dichalcogenides

The history of TMDCs started with the determination of MoS_2 structure by Dickinson and his colleague in 1923[20]. As long ago as 1986, single-layer MoS_2 sheets have been achieved by lithium intercalation[21]. Nevertheless, the rapid growth of TMDCs research contributes to the discovery of graphene monolayer, which powerfully stimulates the studies of ultrathin 2D materials. In particular, the 2D TMDCs attracted increasing attention due to their unique optical and electrical properties for the potential application in the future electronics[12-14]. **Figure 2.1** is the publications citation plot of TMDCs related papers in the past decade that shows the growing tendency of research in TMDCs. In this section, the crystal structure, basic electronic properties and synthetic routes of 2D TMDCs will be briefly discussed.

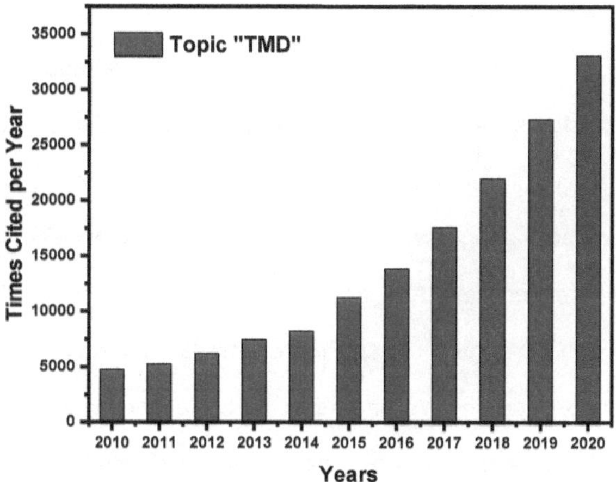

Figure 2.1Publication Citation plot of topic "TMD" in the period of 2010-2020. (Searched by Web of Science database , February 27th, 2021)

2.1.1 Atomic structure of Two-dimensional Transition Metal Dichalcogenides

TMDCs have basic chemical formula as MX_2, where M is transition metal and X is chalcogen elements, for example, molybdenum diselenide ($MoSe_2$), tungsten disulfide (WS_2). The layered structure is composed of one transition metal atomic layer sandwiched by two chalcogen atomic layers, as shown in **Figure 2.2a**.

TMDCs have three-folder symmetry and three atomic thickness for a single layer (around 6-7Å). Among each layer, weak van der Waals interaction could be easily broken even by mechanical exfoliation, allowing to obtain monolayer flakes from bulk materials and the heterostructure formation. **Figure 2.2b** shows the periodic table of transition metal and chalcogen elements which formed at least 40 different TMDCs with layered structure[22]. TMD materials contain two typical phases: trigonal prismatic (hexagonal, H) and octahedral (tetragonal, T), the detailed crystal structures as summarized in **Figure 2.2c**[23]. The H-phase contains the hexagonal crystal symmetry, which is evident from the top view. When there are multilayer stacking together, AA' and AB stacking sequence correspond to 2H and 3R phase, respectively. It is easy to understand that when the layer stacking together with 60-degree rotation represents 2H stacking, conversely, 0-degree represents 3R stacking. When one of the chalcogen layers orderly slides to one direction, T-phase formed, which has the same symmetry in either monolayer or multilayer, called 1T. However, the metal atoms layer distorted or dimerized in one direction would result in distortion T-phase, also called 1T'. The different stacking order of the layers Different phases and stacking orders of TMD structure would result in different electronic and optical properties. In Chapter 4, we synthesized both 2H- and 3R- stacking WSe_2 bilayer, they exhibit different optical properties, and their stacking sequence could be clearly identified by STEM images.

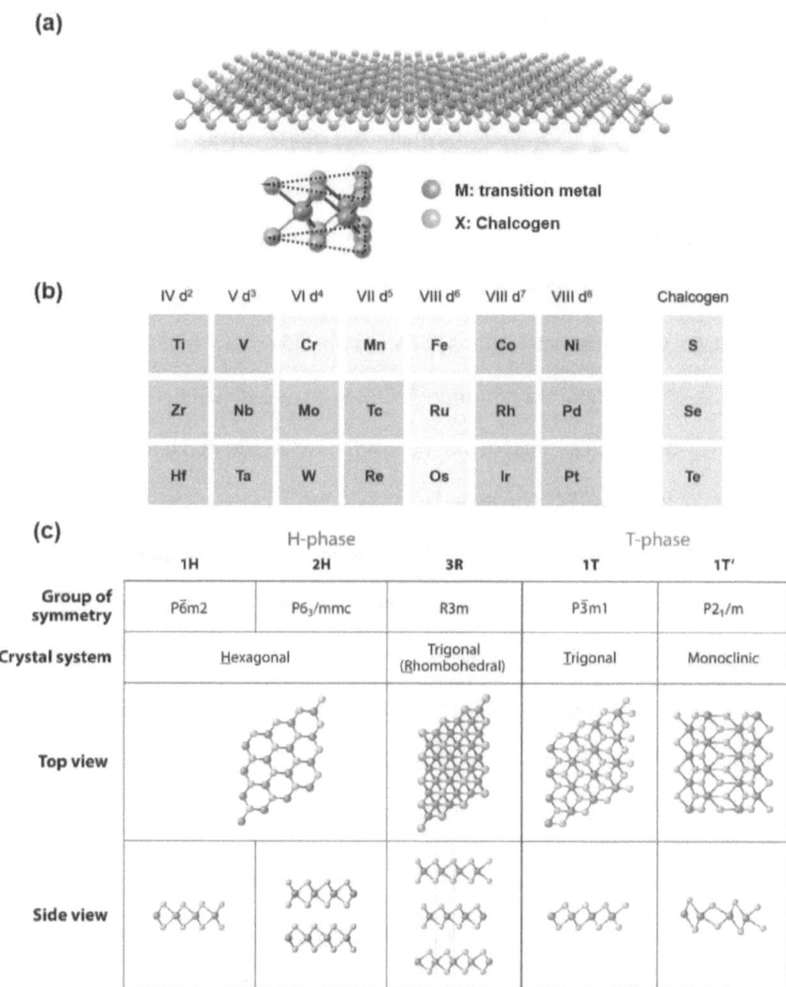

Figure 2.2 Atomic structure of 2D TMDCs. (a) Schematic of typical 2D TMDCs (MX_2). (b) Transition metals table of known layered dichalcogenides (dark blue) and chalcogen elements (yellow). (c) Top view and side view of common TMDCs crystal structures in both H- and T- phases.[23] Blue and purple ball represent` the transition metal atoms (M), and the yellow ball represents the chalcogen atoms (X).

2.1.2 Electronic Properties of 2D Transition Metal Dichalcogenides

In general, most of the TMDCs have both metallic 1T and semiconducting 2H phase, while 2H is the thermodynamically stable phase for TMD materials. Unlike graphene with zero-bandgap, 2D TMD semiconductors with a finite bandgap are tunable from the choice of material, controlling the layer number, strain engineering, and external electrical field. As shown in **Figure 2.3**, the bandgap of TMDCs with different compositions of transition metal and chalcogen cover most of the visible and infrared range[24]. Owing to one dimension has been confined in the quantum scale, most of the monolayer TMDCs reveal direct bandgap, whereas most of the bulk TMDCs are indirect bandgap. For example, the indirect band gaps are simulated to be 1.23 eV, 1.09 eV, 1.32 eV, 1.21 eV for bulk MoS_2, $MoSe_2$, WS_2, WSe_2, respectively[25]. However, when the layer number decrease to one, the estimated result become direct band gap of 1.88 eV (MoS_2), 1.57eV ($MoSe_2$), 2.03eV (WS_2), 1.67eV (WSe_2)[25]. For these monolayer II-VI 2H-TMDCs, the direct bandgap also leads to strong photoluminescence (PL), which creates the potentiality of many electro-optics applications.

Furthermore, the unique electronic band structure of the two inequivalent momentum valleys (K, K') empowers valley polarizability, which is not observed in their bulk counterpart[26]. Moreover, 2D TMDCs have a flat surface free from dangling bonds and defects, resulting in less electron scattering and relatively large charge flow with high mobility. For example, the simulated mobility of WS_2 reaches 320 cm^2/Vs for electrons and 540 cm^2/Vs for holes[27]. Although, there still remain several issues that need to be solved to realize the high mobility, such as the choice of contact metal, back gate oxide and the decrease of defect density and grain boundary. In Chapter 6, we proposed a new approach to grow a high-quality WS_2 monolayer with a low defect density of around 10^{12} cm^{-2}, which promotes electron mobility to 185 cm^2/Vs at room temperature.

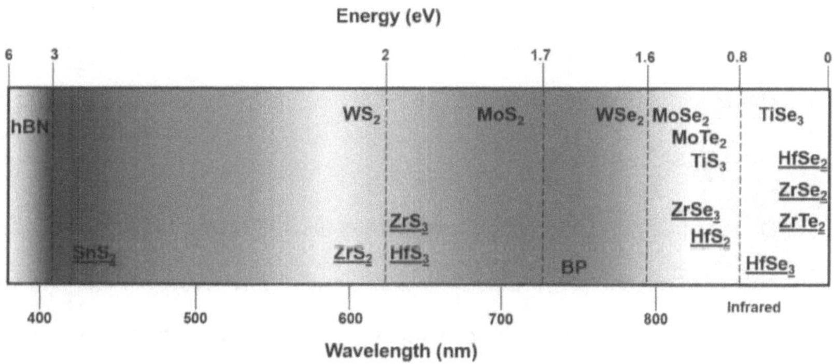

Figure 2.3 Schematic of 2D materials bandgap distribution from infrared to ultraviolet. The color presents the corresponding wavelength of the bandgap. The name of the materials with indirect bandgap has been underlined. Without underline represents a direct bandgap.

2.1.3 Fabrication of ultrathin 2D Transition Metal Dichalcogenides

The pioneer top-down methods rely on the various exfoliation methods of layered materials, including mechanical exfoliation (ME), chemical exfoliation, and laser thinning[28]. **Figure 2.4a** schematically presents the ME technique of thin layers or monolayer TMDCs from their bulk crystal[8]. ME makes single-crystal flakes with high quality favorable for the preliminary study of the properties and investigation of the material for particular topics. Very recently, the most massive centimeter-scale exfoliated monolayer 2D TMDCs have been achieved by Zhu's group[8], where the gold film was applied to strip off the 2D materials layer-by-layer. The monolayer TMD films could transfer to a targeted substrate like SiO_2/Si or other 2D materials to establish the heterostructure (**Figure 2.4a** right panel). Nevertheless, the ME process has exceedingly low scalability due to the challenge of growing large-scale single-crystal bulk materials. In the meantime, the irregular shape and uniformity of TMDCs and chemical residue during the transfer process prevent practical applications.

On the other hand, the bottom-up approaches for TMDCs synthesis include vapor-phase deposition, hydrothermal, and thermolysis synthesis (solution-phase). Among these, chemical vapor deposition (CVD) is able to produce high-quality TMDCs thin sheets with a large area, manageable thickness, and good physical properties, which are the essential requirements for their practical use in electronic and electro-optical devices. To date, the large-area synthesis of ultrathin TMDCs coating using the CVD method has been demonstrated by several research groups. In the pioneer experimental report by Li's group[6], sulfur (S) and molybdenum trioxide (MoO_3) powders were separately placed in the proper locations of the reaction chamber (**Figure 2.4b**). Then, the target substrates were placed above the MoO_3 container. During the chemical vapor growth, MoO_3 and S are thermal evaporated and then undergo a two-step reaction ($MoO_3 + x/2S \rightarrow MoO_{3-x} + x/2SO_2$ and $MoO_{3-x} + (7-x)/2S \rightarrow MoS_2 + (3-x)/2SO_2$). MoO_{3-x} is an intermediate phase during the reaction and then diffuses to the surface of the target substrate. Thereupon, MoO_{3-x} further reacts with sulfur vapors to form MoS_2. The approach is capable of achieving the single-crystalline various TMD thin flakes or thin films directly on arbitrary substrates via proper nucleation density control. **Figure 2.4b** presents the general setup components of CVD methods; meanwhile, the photo on the right side shows the quantity production of MoS_2 thin film through CVD[29]. CVD method is universal for fabricating most kinds of TMD thin layers and their heterostructures, for example, $MoSe_2$, WS_2, WSe_2, NbS_2, PtS_2.

One simply modified CVD method is metal-organic chemical vapor deposition (MOCVD), which using a metal-organic source to provide vapor phase precursor into the reaction chamber (**Figure 2.4c**). Wafer-scale monolayer MoS_2 and WS_2 have been achieved by Park's group[16] via MOCVD with molybdenum hexacarbonyl $[Mo(CO)_6]$, tungsten hexacarbonyl $[W(CO)_6]$, and diethyl sulfide $[(C_2H_5)_2S]$ as precursors. As grown monolayer MoS_2 samples show average electron mobility as high as 30 $cm^2V^{-1}s^{-1}$. This work shows significant potential for a practically scalable technique for TMDCs thin film since it gives a better route to precise control over precursors of metal and chalcogen during synthesis. More detailed mechanisms and modifications of CVD will be detailed discussing in the following section.

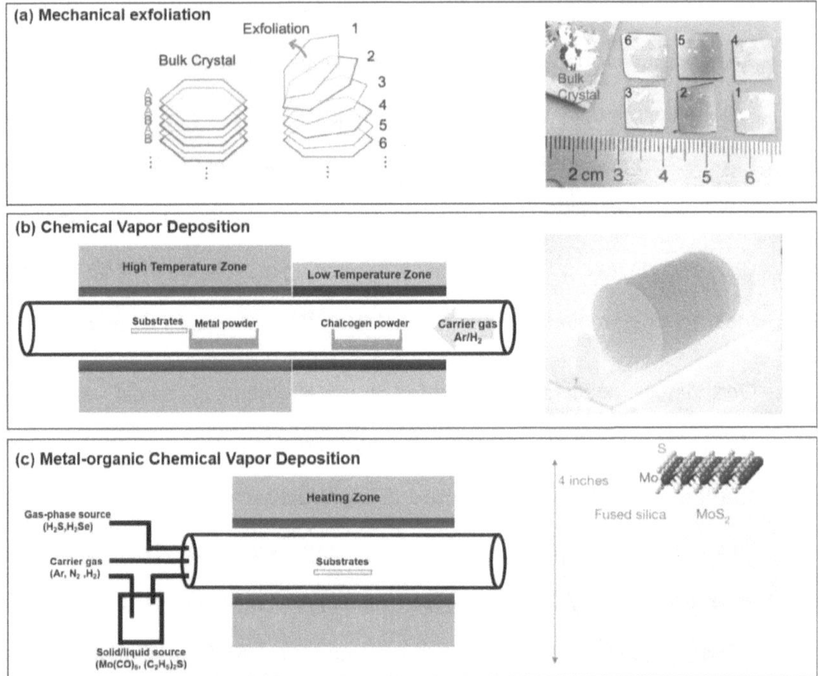

Figure 2.4 Fabrication techniques for ultrathin 2D TMDCs. (a) Schematic illustration of layer-by-layer ME of 2D TMDCs sheets. The right panel is the image of centimeter-scale monolayer WSe$_2$ on the SiO$_2$/Si substrates by ME approach[8]. (b) Schematic illustration of CVD setup. The right photo is the MoS$_2$ monolayer on a 2-inch sapphire wafer[29]. (c) Schematic illustration of MOCVD setup. Photo of 4-in MoS$_2$ wafer achieved by MOCVD method on the right side[16].

2.2 Layer Controllability of Chemical Vapor Deposition

Few-layers TMDCs films exhibit different electrical and optical properties to their monolayer structure, which enable a broad range of applications such as field-effect/tunneling transistors[30], chemical sensors[31], and photovoltaic applications[32].

However, TMDCs materials prefer the layer-plus-island growth mode in the CVD system, leading to an uneven surface coverage with meager yield for multilayer crystals. In this Section, the fundamentals and several proposed mechanisms of layer controllable CVD growth will be introduced.

2.2.1 Fundamentals of van der Waals epitaxy growth

The nature of 2D materials with a free dangling bond leads to a unique situation for epitaxy, different from conventional epitaxy, van der Waals (vdW) gaps instead of strong chemical bonding formed between epitaxy 2D layers the substrate, so-called van der Waals epitaxy growth. Although the vdW gaps benefit the misfit lattice growth, the nucleation and orientation of the epitaxy film are governed by the substrate. The lattice mismatch and the thermal expansion between the vdW film and substrate would also introduce a critical amount of strain. **Figure 2.5a** illustrates the deposition model for vdW epitaxy growth. Series steps happened on the substrate surface, including seed formation, adsorption/desorption of precursor atoms, adsorbed atoms diffusion, and growth of vdW film. Moreover, these deposition steps are strongly related to the CVD growth environment variables such as temperature, pressure, and gas flux. Therefore, the mastery of fundamental mechanisms toward nucleation and growth on the substrate surface would be beneficial to control the appropriate growth condition to get desired 2D materials.

The nucleation happens when the precursor concentration is supersaturated on the substrate surface; in other situations, artificially created seeds and big particles could also act as a heterogeneous nucleation site. There are two proposed nucleation sites in a conventional CVD system for 2D TMDCs due to the precursor ratio variation. Take CVD growth of MoS_2 as an example, MoO_3 precursor is reduced and reacted with deficient S vapor to form sulfurized molybdenum oxide, which drops into the substrate as nucleation site (**Figure 2.5b**)[33]. Due to the chemical bonding between these oxide particles with the substrate, relatively large strain or wrinkles would happen in the resulted MoS_2 flakes. On the other hand, sufficient S vapor reacts with MoO_3, which supplies supersaturated MoS_2 vapor and nucleate on the substrate surface (**Figure 2.5c**) [34]. MoS_2 seeds will follow the vdW epitaxy growth mechanism and better for growth control. In Chapter 4, we utilized the feature of concentration variation in the CVD chamber and seed difference to controllable synthesize the monolayer WSe_2 with variable strain level.

There are three basic nucleation modes for thin-film deposition were summarized in **Figure 2.6**[35]. Volmer-Weber growth mode happens when the seeds or spices prefer to stack together rather than touch substrate, leading to multilayer islands or particle growth, also known as island growth mode (Figure 2.6a). Conversely, seeding on the substrate with lower energy than stacking together results in the layer-by-layer growth mode or Frank-van der Merwe growth mode (Figure 2.6b). In which, the subsequent layer will not grow until the first layer is ultimately growth, which is ideally for homogeneous and layer controllable growth. However, most of the 2D TMDCs thin film growth follows a combination growth mode called layer-plus-island growth mode (Figure 2.6c), also known as Stranski-Krastonov growth mode island growth happened after the first layer thin film formation. Therefore, growth mode transformation of TMDCs thin film from layer-plus-island to layer-by-layer is significant to the layer number controllability. In Chapter 4, we proposed a strain-directed growth mode transformation for vdW homo- and hetero-stacking structure.

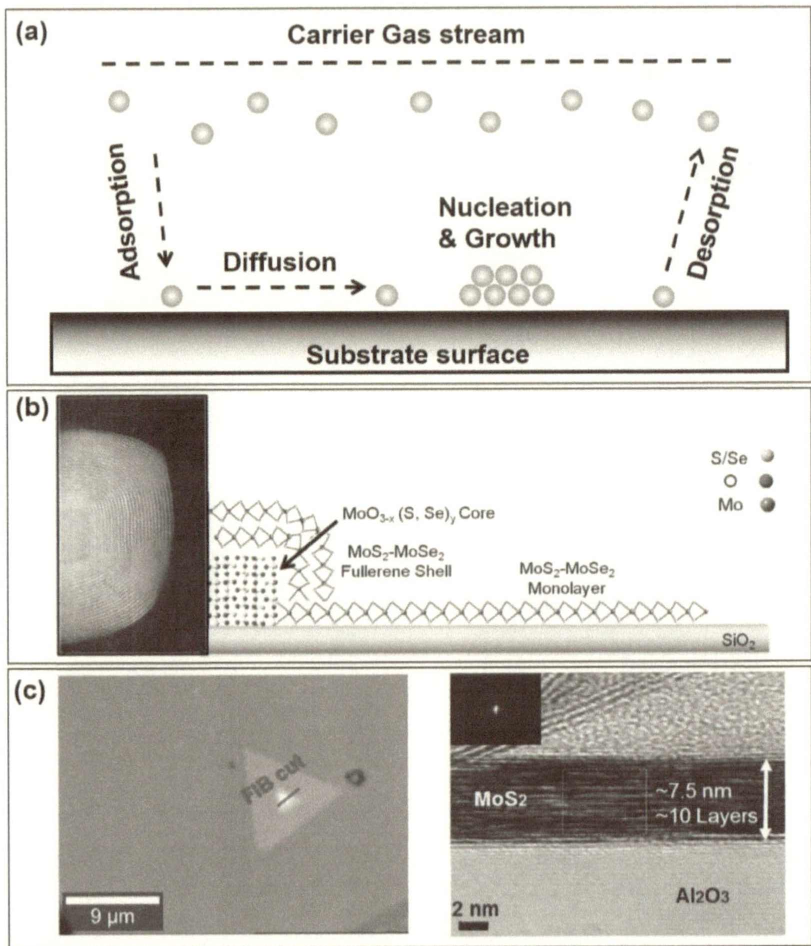

Figure 2.5 Nucleation and growth of van der Waals epitaxy. (a) Schematics of deposition process on the substrate surface for vdW epitaxy growth. (b) Cross-section view of sulfurized molybdenum oxide seed[33]. (c) Optical and TEM cross-section images of MoS$_2$ seeds[34].

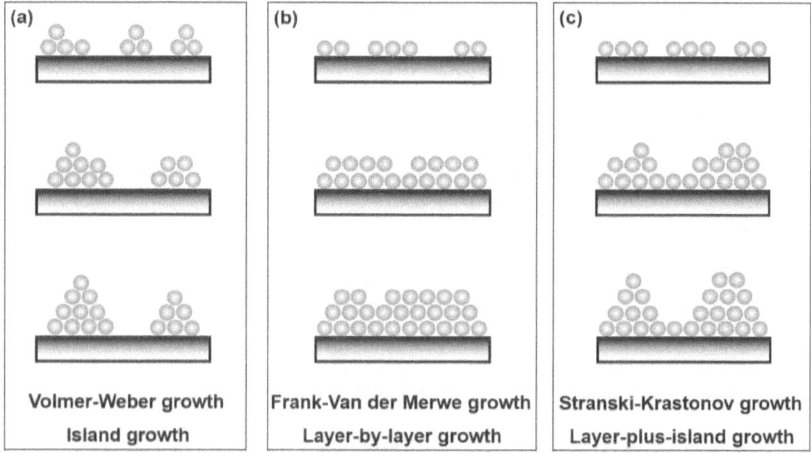

Figure 2.6 Schematic illustration of three basic thin film growth modes. (a) Volmer-Weber growth mode or island growth mode. (b) Frank-Van der Merwe growth mode or layer-by-layer growth mode. (c) Stranski-Krastonov growth mode or layer-plus-island growth mode.

2.2.2 Mechanism study of vertical van der Waals stacking

The discussion about deposition process in the last section shows that the competition between nucleation and growth determines the growth behavior and the final structure of 2D TMDCs materials. Therefore, understanding the kinetics and thermodynamics behavior in the CVD system is critical for growth control. Recently, a few simulation works have been published to study the vertical versus lateral growth mechanism. The integrated density functional theory (DFT) was applied to calculate the formation energy of mono- and bi-layer MoS_2 under various conditions. Shang and his colleagues[36] found that the stability of MoS_2 bilayer structure is size-dependent, larger size bilayers are more stable, while the critical size is related to the substrate choice. Coincidentally, Ye et.al[18] have a similar result that the size of the first layer influence the second layer growth behavior. Based on the thermodynamic analysis, the criterion for second layer growth was expressed as

$$\Delta\varepsilon_{bind} = \varepsilon_{LL} - \varepsilon_{L1S} > \eta(\frac{\gamma_2}{L_2} - \frac{\gamma_1}{L_1})$$

where η is the structural parameter, γ is the edge energy, L is the edge length, ε_{LL} is vdW binding energy density between 2D materials and ε_{L1S} is the vdW binding energy density between the first layer and the substrate. The energy difference $\Delta\varepsilon_{bind}$ should large enough to overcome the energy penalty of second layer growth (**Figure 2.7a**). Thus, there exists the critical size of L1 for the second layer growth.

Besides the formation energy of 2D TMDCs, the kinetic properties of the precursor atoms have a significant impact on the growth result. In a conventional CVD system for MoS_2 growth, the diffusivity of the S precursor is much faster than that of Mo, thus the composition and diffusivity of Mo precursor control the growth behavior[36]. The temperature, pressure, flow rate, and substrate will influence the diffusion properties in the growth chamber. The temperature would modify the diffusion coefficient and kinetic coefficient, therefore, there is a critical temperature to initial the bilayer growth, which is simulated as shown in **Figure 2.7b**[18]. The higher temperature also supplies enough thermal energy for overcoming the energy barrier for multilayer nucleation. The researcher also found that the flow rate of carrier gas modifies the vertical growth behavior. Under relatively more minor flux, the size of the second layer is more prominent, as shown in **Figure 2.7c**[18]. Most of the theoretical study of layer-dependence growth analyzed the growth mode kinetically and thermodynamically. Seldom of them discuss the influence from the substrate and other properties of the first layer like strain. In chapter 4, the DFT calculation has been developed to discuss the model of strain effect in vertical growth.

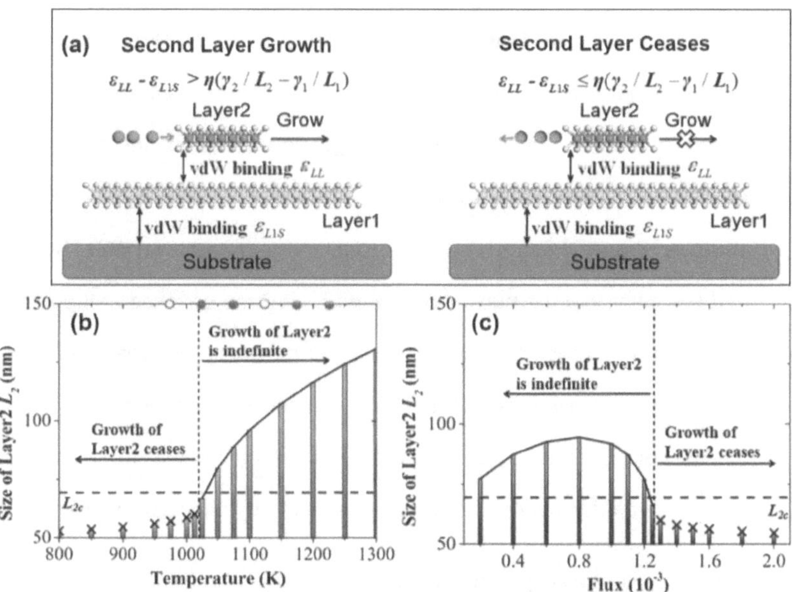

Figure 2.7 Thermodynamic and kinetic study for vertical vdW stacking[18]**.** (a) Schematic illustration of thermodynamic criterion for vertical vdW growth on the substrate surface. (b) Temperature-dependent size variation of the second layer, the dashed line indicates the critical temperature for second layer growth. (c) Flux-dependent size variation of the second layer, the dashed line indicates the cease of second layer growth.

2.2.3 Synthesis of van der Waals Homostructure

In a conventional CVD system, after the growth of the monolayer TMDCs, random multilayer islands start to grow on top of the first epilayer, known as layer-plus-island growth mode, as shown in **Figure 2.8a,** which is hard to get homogeneous multilayer crystals. Therefore, most of the studies investigated the synthesis of single-crystal bilayer TMD islands (seeds). In the last section, we mentioned that the seeding of the second layer starts from the center of the crystal and enlarges during the growth window. A new growth principle for the TMD homostructure was recently proposed by Fang et al., as shown in **Figure 2.8b and c**[37]. The seed of the second layer started from the edge and center of the monolayer flakes simultaneously and continuously grew until they merged to form a uniform bilayer structure. **Figure 2.8d** presents the optical image of as-grown WSe_2 homo-junction, and it is clear that the nucleation started from three vertexes of the monolayer[37]. This edge nucleation has also been found in vertical heterostructures growth[38], which will be detailed discussing in the next section. The mechanism proposed in the last section could explain most of the experiment results toward the controlled growth of van der Waals homostructures thermodynamically and kinetically.

Kinetically, the raising of the growth temperature promotes the vertical growth for the homostructure. As shown in **Table 2.1**, we summarize the growth temperature of CVD grown MoS_2, $MoSe_2$, and WSe_2 for their monolayer (1L) and bilayer (2L) structures. For molybdenum-based TMDC materials, the growth temperature is relatively lower than that of tungsten-based TMDC because of the lower melting point of MoO_3 precursor compared to WO_3 precursor. Monolayer MoS_2 crystals were achieved at 700~800°C, and the growth temperature of bilayer MoS_2 flakes is about 50-100°C higher than that for 1L[18, 39, 40]. Similarly, bilayer structure of $MoSe_2$, WS_2 and WSe_2 crystals were obtained with increasing the growth temperature of 75-100°C compared to the monolayer growth condition[18, 41, 42]. Moreover, the continuously elevated growth temperature to > 900°C leads to few layers (e.g. 3, 4 layers) growth. Fang et al[37] demonstrated that the yield of the bilayer crystal also strongly related to the growth temperature, their bilayer WSe_2 yielded 30% at 880°C, which increased to 80% at 940°C. On the other hand, the growth temperature of the vertical stacked homostrucutre has greatly impact on the stacking sequence for the multilayer crystals. The simulation work by He et al[43] proposed that the AA' pattern is the most stable stacking order, however, there is only 5 meV per formula unit difference between AA' and AB stacking pattern. Therefore, the growth temperature become critical to separate the AA' and AB stacking sequence. At 850°C, a mixture of AA' and AB stacking bilayer MoS_2 were achieved, and AB stacking dominated by increasing the growth temperature to 900°C[39, 44]. An absolutely separation of AA' and AB stacking MoS_2 were demonstrated by a two-step CVD process with reverse flow of carrier gas, which drastically decreased the nucleation during the ramping process[40].

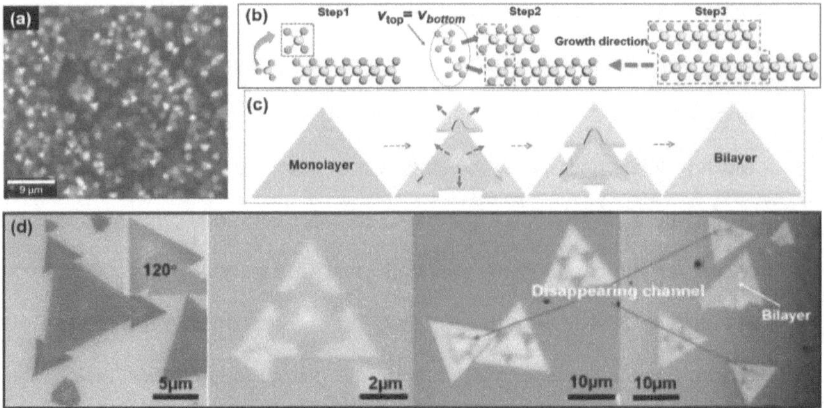

Figure 2.8 Center and edge nucleation for van der Waals homostructure. (a) Optical image shows the multilayer WSe_2 due to the layer-plus-island growth mode; the background color corresponds to 1L. (b) Schematic illustration of the edge nucleation and growth process for the second layer. (c) Schematic illustration of the edge and center nucleation to get fully covered bilayer structure. (d) Optical images illustrate the WSe_2 homo-junction at different growth stages. (b, c, d) from ref[37].

On the other hand, the precursor concentration on the substrate surface also modulates the epitaxial growth mode and speed. For example, the adjustment of gas flow rates affected both transition metal and chalcogen precursor, while the vertical epitaxy happened at a low flux rate with a high concentration on the sueface[18]. To increase the transition metal concentration, transition metal chloride like $MoCl_6$ with a low melting point replaced the oxide precursor, and the increased amount of $MoCl_6$ lead to the bilayer MoS_2 film growth[45]. Some works used salt like sodium chloride (NaCl) to promote the metal oxide delivery and achieved multilayer TMD flakes[46]. Multilayer MoS_2 (up to 20 layers) have been synthesized using oxidized Mo foil as the precursor, which the layer number can be controlled by tuning the oxidation degree of Mo foil[47].

Meanwhile, the reductant like hydrogen gas reduced the oxide precursor promoting the transition metal precursor delivery; however, too much hydrogen gas will also etch the TMD growth. Fang and his colleagues detailed discussed the effect of hydrogen gas in the growth of MoS_2 monolayer and bilayer structure[48]. The simplest way to have a condensed concentration of transition metal precursor is to deposit thicker metal oxide on the substrates and sulfurized or selenized it under high temperature[49]. Except to control the transition metal precursor, the chalcogen precursor amount will also modify the growth mode. Bilayer even trilayer MoS_2 and WSe_2 samples have been attained by increasing the temperature or amount of S or Se powders[50, 51].

Thermodynamically, the substrate surface energy and interaction between the substrate and TMD significantly impact the growth results. Jeon et al. applied different duration of O_2 plasma treatment with SiO_2 substrates, which give rise to the controllable growth of 1-3 layers MoS_2 film[52]. After the O_2 plasma treatment, there existed Si-(O or OH)$_4$ bonding which provides a higher surface reactivity for the multilayer TMD growth[52]. Chen et al. demonstrated the terrace on the sapphire substrate as a better nucleation site could modulate the growth mode to the layer-over-layer and induce the aligned bilayer WSe_2 growth[53].

These works of literature show many fundamental and interesting mechanisms and methods to the controllable synthesis of the vertical homostructures TMD.

However, less of them notice the growth mode transition between the monolayer and subsequent layers growth, which is critical to have wafer-scale controllability of multilayer TMDCs. Meanwhile, it is still challenging to have large-scale multilayer (>20L) TMD growth with homogeneous layer numbers. These multilayer structures with high mobility could also be applied for many industrial applications different to the monolayer structure. Therefore, it is demanded to devote more effort into the layer controllability engineering in the CVD system for the 2D TMD materials.

Table 2.1 Summarization of CVD growth temperature for monolayer (1L) and bilayer (2L) TMDCs.

Materials	T for 1L	T for 2L	Comments and Reference
MoS_2	800°C	900°C	[39] AB stacking dominate
MoS_2	-	850°C	[44] 80% yield, mixed AA' and AB stacking
MoS_2	730°C	730°C	[54] 10 minutes for 1L, 15 minutes for 2L
MoS_2	700°C	750-800°C	[40] AB stacking bilayer at 750°C , and AA' stacking bilayer at 800°C
MoS_2/WS_2	700°C	800°C	[18] low coverage of 2L at 750°C
$MoSe_2$	750°C	825°C	[41] 3-4 layers were achieved at 900°C
$MoSe_2$	700°C	750°C	[18]
WSe_2	850-950°C	950-1050°C	[42] few layers samples also exist
WSe_2	-	880-940°C	[37] yield 30% for 880°C, 60% for 900°C, 80% for 940°C
WSe_2	850°C	950°C	[18] low coverage of 2L at 900°C

2.2.4 Synthesis of van der Waals Heterostructure

The van der Waals heterostructure synthesis consists of lateral and vertical growth, as shown in Figure 2.9. For the lateral heterostructure, one kind of TMD layer grows first on the substrate, and another kind of TMD layer grows from the edge of the first layer due to the low binding energy at the edge. Gong and et al.[55]

first demonstrated the MoS$_2$/WS$_2$ lateral and vertical heterostructures using a one-spot CVD system. Mo source has a lower melting point; thus, the MoS$_2$ grows first, and WS$_2$ grows at the outside at 650°C. The vertical stacking could be realized by elevating the growth temperature to 850°C[55]. To better control the delivery sequence of W/Mo precursor in a one-spot CVD system, core-shell WO$_{3-x}$/MoO$_{3-x}$ nanowires have been used as the precursor for the vertical stacked MoS$_2$/WS$_2$ heterostructures[56]. Like the homostructure synthesis, the growth mode of heterostructures is strongly related to the growth temperature and precursor concentration. Higher growth temperature and higher precursor concentration lead to a vertical stacking preference, which also induces an alloy formation in the hetero system, especially at the interface between two materials.

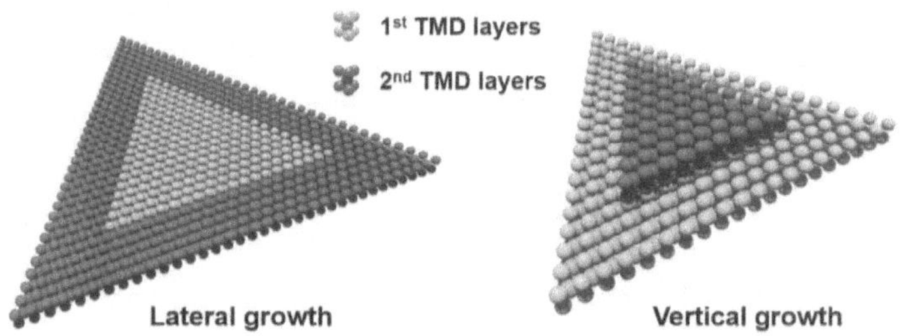

Figure 2.9 Schematic illustration of Lateral and vertical growth of van der Waals heterostructures.

Therefore, the two-step CVD method was applied to grow each material separately and achieve a clean interface between the two materials. Li et al. [57] have demonstrated a clean interface between MoS$_2$ and WSe$_2$ heterostructure, which is hard to obtain by a one-spot CVD system. Meanwhile, the two-step CVD system has better control and understand for the growth mechanism study. For example, Yoo et al. [58] found that the contamination and defect density of the first epilayer MoS_2 has dramatically impacted the second material WS_2 growth behavior. The MoS_2 monolayer with small particles on top induced vertical bilayer MoS_2/WS_2 heterostructure, conversely, clean MoS_2 monolayer direct the lateral heterostructure.

On the other hand, the exact first epilayer under different growth conditions at the second time would have different growth behavior. Simply, the higher growth temperature manipulates the vertical bilayer heterostructures in the second step due to the kinetically preference[59]. In the second TMD growth system, the rational control of metal/chalcogenide precursor could also direct the heteroepitaxy laterally or vertically. Li and his colleagues have discussed the detailed mechanism, that the nucleation starts from the edge of the first layer similarly to the vertical homoepitaxy growth[38]. Various heterobilayer structures such as MoS_2/WSe_2, $WS_2/MoSe_2$, WS_2/NbS_2, and MoS_2/NbS_2 have been successfully achieved.

By better control the delivery of different transition metal sources, superlattice structure could be obtained by CVD systems[17, 55, 60, 61]. Zhang et al. demonstrated the TMD mutiheterostructures and superlattices using a step-by-step CVD process with a reverse flow which avoids the nucleation during the ramping process[17]. A similar structure has also been obtained by a water-assisted one-spot CVD system, which controls the delivery of metal sources by changing the carrier gas[60]. This setup also demonstrated a bilayer MoS_2-WS_2 lateral heterostructure with a clean interface[55].

2.3 Strain and Defect Engineering on 2D Transition Metal Dichalcogenides

The mechanical flexibility of 2D TMDCs offers an excellent platform for strain engineering. Atomically speak, strain changes the materials' atomic configurations in one direction or all directions, manipulating the electronic and optical properties of 2D TMDCs[62, 63]. Meanwhile, defects like vacancies or dopants also change the specific atomic structure locally or periodically, which has been demonstrated a significant impact on the electrical performance and optical properties of 2D TMDCs[64]. In this chapter, we will discuss the kinds of strain and defect that have been observed in 2D TMDCs. The related properties variation will also be briefly introduced. More important, we will discuss the CVD synthesis engineering for the strain and defect controllable growth. Specially, the defect engineering discussed in this chapter is for the high-quality TMDCs samples with low defect density.

2.3.1 Strain Type and Level on 2D TMDCs

In structures of TMDCs, the mechanical properties could be different along with the armchair and zigzag direction. Therefore, the strength applies to different directions yielding different strain types, including uniaxial and biaxial strain. By stretching or compressing the TMDCs, the strain could be sorted as tensile strain and compressive strain. Theoretically, the ultimate strain that MoS_2 can afford is

0.24, 0.37, and 0.26, along armchair, zigzag, and biaxial deformation[65]. Bertolazzi et al. have experimentally demonstrated that the upper limit breaking strength of monolayer MoS_2 is 11%[66]. Strain also exists out of plane. The out of plane deformation including rippled, wrinkled, and crumpled, contains complicated strain distribution where the tensile and compressive strain exist simultaneously.

The strain engineering in 2D TMDCs can tune both the electrical and optical properties. Therefore, the optical properties variation like the shift of photoluminescence (PL) emission peak or adsorption peak can help to determine the strain level in TMDCs materials. Experimentally, the tensile strain has a linear relationship to the redshift of PL and absorption peaks. For monolayer MoS_2, the adsorption peak presented a redshift rate of 64±5 and 68±5 meV/(% of uniaxial tensile strain) for the A and B exciton, respectively[67]; and the PL emission peak presented a redshift rate of 48 meV/ (% of uniaxial tensile strain) for the A exciton[68].

For monolayer WSe_2, the redshift rate of 54±2 meV/ (% of uniaxial tensile strain) was derived for the A exciton[69]. In addition, Frisenda and et al. have theoretically and experimentally studied the relationship between biaxial strain and the bandgap transition[70]. They concluded the gauge factors of the different TMDCs with a sequence that $MoSe_2$ (-33meV/%)< MoS_2 (-51meV/%)< WSe_2 (-63meV/%)< WS_2 (-94meV/%) for the A exciton[70].

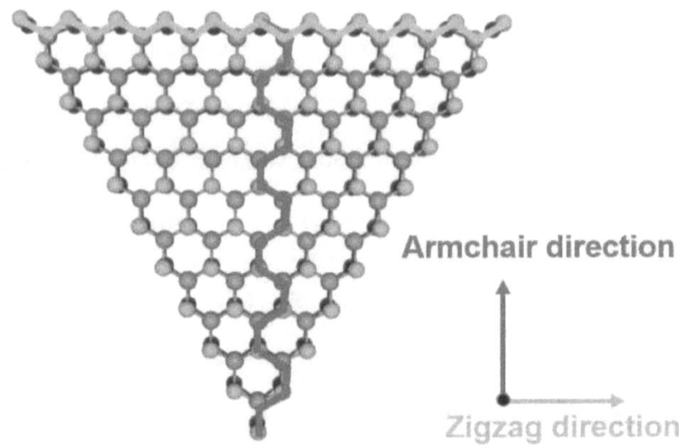

Figure 2.10 Schematic illustration of the orientations in 2D TMDCs. The red line presents the armchair direction and the green line presents the zigzag direction.

2.3.2 Strain Engineering on 2D TMDC by Chemical Vapor Deposition

The most well-known approach to controllable applied tensile strain to a 2D TMDCs is transferring the TMDCs thin layers to flexible substrates like PMMA and bending them. However, this approach could only achieve uniaxial tensile strain and hard to apply to a large scale. Therefore, it is critical to study the large-scale and strain controllable CVD method for 2D TMDCs. This section briefly introduces the strain formation approaches during van der Waals epitaxial growth.

Substrates play a significant role during epitaxial growth. The difference of lattice constants and the thermal expansion coefficients (TEC) between substrates and 2D TMDCs generates the in-plane localized strain during the CVD growth. As illustrated in **Figure 2.11**, when the TEC of the substrate smaller than that of the TMDCs materials, the substrate has a slight lattice variation during the quenching process, which stretches the TMDCs materials and generates biaxial tensile strain on 2D TMDCs. Conversely, biaxial compressive strain build on 2D TMDCs when the TEC of substrate larger than that of TMDCs. Regardless of lattice constants, free strain is induced when the TEC is equal between substrates and TMDCs. Ahn and et al.[71] have demonstrated strain-engineered growth of WSe_2 on substrates with different TEC. Up to 1% biaxial tensile strain and 0.2% biaxial compressive strain has been achieved by silica substrate and strontium titanate substrate, respectively[71].

On the other hand, TMDCs thin layer strain level can be manipulated by the nanostructure fabrication on the substrate surface. The simplest way is to pattern the substrate first and then transfer the 2D TMDCs onto the patterned substrate to induce strains. The patterned structure could be designed as nanopillars, nanoholes, and nanorods to get the localized strain by bending or stretching the 2D film[62]. Direct grown strained TMDCs layers on patterned substrates to get large-scale strained TMDCs materials were realized early in 2014. Tan and et al.[72] used the CVD method to grow monolayer MoS_2 on nanoporous gold generating out-of-plane strains. Besides, the strained MoS_2 grown on patterned nanocone and nanopyramid SiO_2 substrates has been demonstrated by deposition MoO_3 precursor and sulfurization[73].

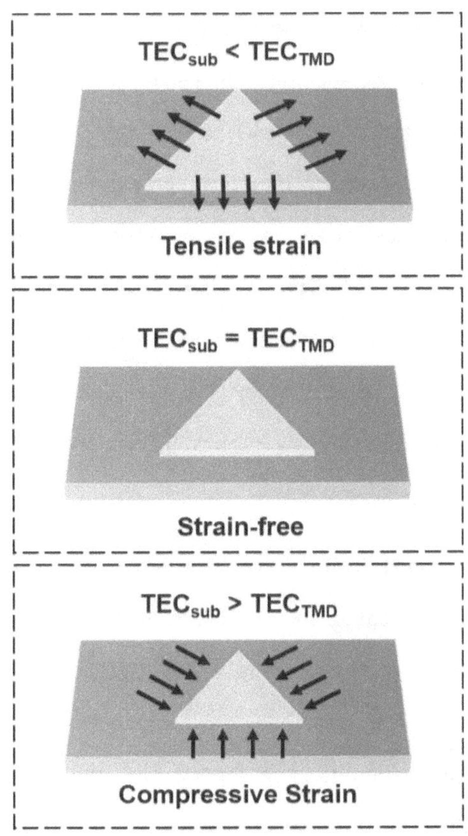

Figure 2.11 Schematic illustration of biaxial strain generated from the thermal expansion coefficient difference between substrate and TMDCs layers.

The strain also exists in the TMDCs heterostructures due to the lattice mismatch. For the vertical heteroepitaxy, the strain could be relaxed without dislocation formation due to the special van der Waals gap. The inhomogeneous strain has been found in the WSe$_2$-MoS$_2$ lateral heterostructure due to their lattice mismatch[74]. Furthermore, these in-plane strain may be released when misfit dislocation[75] or defect (alloy) is formed in the interface. Xia and et al. demonstrated that the relative width between WS_2/WSe_2 superlattice manipulated the strain level and ripples formation[61].

The defect and doping effect could change the atomic structure of 2D TMDCs. Thus, the strain would be generated when a large amount of defect or dopants are introduced. Azcatl and et al.[76] have proved that nitrogen-doping can induce compressive strain in MoS_2 structure. Their simulation results exhibited that the strain level was manipulated by the doping concentration, which the N_2 plasma process could control.

2.3.3 Atomic Defects in 2D TMDCs

According to the classification of crystallographic defects, the atomic defect types in 2D TMDCs can be summarized as point defects and line defects, as shown in **Figure 2.12**[77]. The point defects are the most abundant in TMDCs, including vacancies, anti-sites, and substitutions. The vacancies and anti-sites are the most straightforward defects, which do not require other source doping. Take MoS_2 as an example, vacancies contain single sulfur vacancies (V_S), double sulfur vacancies (V_{2S}), molybdenum vacancies (V_{Mo}), and bonded vacancies (like V_{MoS3} or V_{MoS6}), while anti-sites could be a molybdenum atom occupying S_2 column or S_2 column occupying molybdenum position[78]. The substitution defects happen when a foreign atom occupies the intrinsic elements. This foreign atom should follow similar properties as the intrinsic elements like atomic size, electronegativity as the original element. In TMDCs, the replacement between tungsten and molybdenum atoms, and the replacement between sulfur and selenium atoms are common. The monolayer $Mo_{1-x}W_xS_2$ alloys have been investigated during the CVD growth for tuning the bandgap of materials[79].

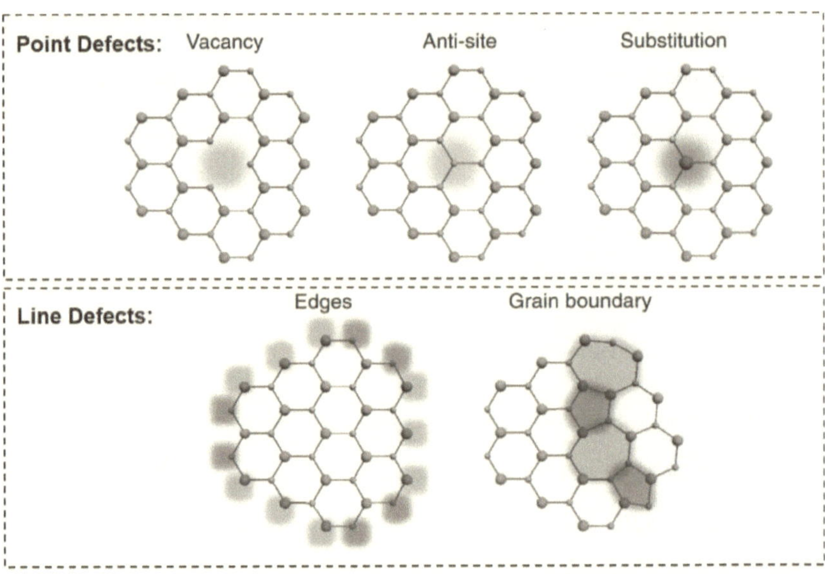

Figure 2.12 Atomic defect types in 2D TMDCs. (Upperside) Point defects include vacancies, anti-sites, and substitutions. (Downside) Line defects include edges and grain boundaries.[77]

The edges and grain boundaries are the commonly found line defects in 2D TMDCs (**Figure 2.12**). The edges of TMDCs could be separated as transition metal terminated and chalcogen terminated edges. Since the chalcogen-rich environment in the CVD system, the as-grown TMDCs have a triangle structure with chalcogen terminated edges. By controlling the precursor concentration, a hexagonal structure with both transition metal terminated edges and chalcogen terminated edges was obtained[80]. Interestingly, the dendritic monolayer MoS_2 with plenty of edges defect was synthesized for electrocatalysis application[81]. The grain boundaries occur at the interface of two grains with different orientations, which significantly limits the charge flow in the 2D TMDCs device. However, the grain boundaries commonly exist in the large-scale thin film by CVD methods. Therefore, how to grow a well-aligned single-crystal wafer-scale thin film has attracted much attention. It should be emphasized that the TMDCs with mirror orientation (0 and 60 degrees) still formed grain boundaries, unlike graphene[78]. Meanwhile, the high strain effect in 2D TMDCs also induces grain boundary formation because of the atomic structure variation.

In most scientific reports, chalcogen vacancies (like V_S and V_{2S}) are considered the most abundant defects due to their low formation energy[78, 82-84]. Very recently, Barja and et al. have demonstrated that the oxygen substituting sulfur defects (O_S) are the dominant defect type in vapor-phase epitaxial TMDCs materials, rather than vacancies[85]. They found that the dominant O_S defect has no in-gap state, which could be identified by scanning tunneling microscopy (STM) techniques. Another well-known technique to observe these atomic-scale defects was the high resolution transmission electron microscopy (TEM). The advantage and disadvantages of both characterizations will be discussed in Chapter 3.

2.3.4 Defect Engineering in 2D TMDCs

The approaches to generate defects in TMDCs and their applications have been detailed reviewed[64, 77]. This section will briefly discuss the reported approaches for reducing the defect density to get high optical and electrical quality 2D TMDCs materials.

In general, the samples from mechanical exfoliation exhibit higher quality (lower defect density) than the CVD-grown TMDCs. The main reason is that the bulk materials for exfoliation are usually fabricated by the chemical vapor transport (CVT) method, which sealed fixed ratio precursors in a high vacuum tube with a prolonged cooling process (recrystallization). The defect density of CVT-grown TMDCs is around 10^{12} -10^{13} cm^{-2}, where the dominant defect type is point defect, like transition metal vancies[83, 86]. These point defects introduce the localized state within the band gap and slow the electron transport. Meanwhile, the chalcogen vacancies lead to an n-doping semiconductor, and the transition metal vacancies lead to a p-doping semiconductor. Yu and et al. have reported that sulfur vacancies could be repaired by thiol chemistry. The improved high-quality MoS_2 field-effect transistors presented high carrier mobility to 80 cm^{2}V^{-1}s^{-1} at room temperature[87]. Furthermore, Edelberg and et al. have reduced the total defect density below 10^{11} cm^{-2} in TMDCs bulk materials by the self-flux growth method in which a sufficient amount of chalcogen is sealed into the tube to form a chalcogen flux[86].

Unlike the CVT or self-flux process, the gas flow is dynamic in the CVD process, which results in the precursor ratio is variable during the growth process.

Simultaneously, the CVD growth chamber's pressure is relatively high (1-760 torr), while the gas leakage may readily happen and induce external dopants like oxygen. As a result, the defect density is > 10^{13} cm^{-2} in CVD-grown TMDCs larger than the ME samples. Kastl and et al. have demonstrated the high-quality WS_2 monolayer by ultra-clean CVD process where the metal oxide source was separated from the gaseous chalcogen source[88]. The as-grown monolayer WS_2 flakes exhibited a lower defect density as 10^{12} cm^{-2}. In addition, the defect control by metal-organic chemical vapor deposition process was investigated, and the as-grown WS_2 reached a low defect density around 10^{12} cm^{-2}. The researcher manipulated the defect density by the post-annealing of WSe_2 film under a chalcogen gaseous environment for ten minutes, which dramatically repaired the Se vacancies and reduced the small oxide particles[89].

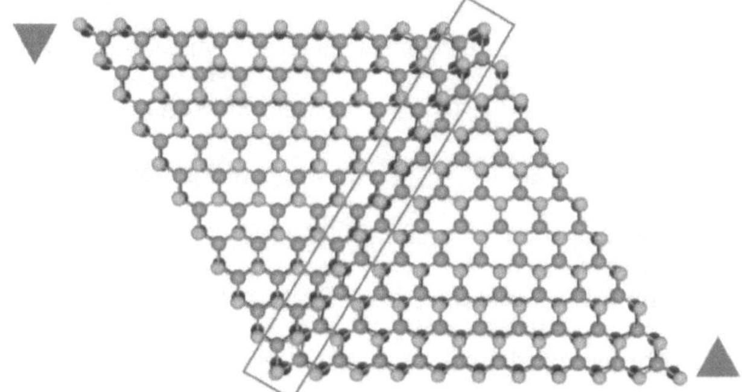

Figure 2.13 Schematic illustration of 60° grain boundaries in TMDCs materials.

The grain boundaries are the other dominant defects in TMDCs materials synthesized by van der Waals epitaxy growth, especially for the large-scale TMDCs film. However, to realize the practical application of these 2D semiconductors, the wafer-scale quality is critical. Since the unique 2D structure, the grain boundary would be formed when two grains with different orientations merged. To solve the issue, scientists were trying to control the orientations of the nucleation and grains. Early in 2015, Chen and et al. found that the sapphires' step edge could guide the WSe_2 growth orientation[53]. And our group members also found that the sulfur concentration during the seeding period has a significant impact on the MoS_2 growth orientation[90]. The well-aligned triangle monolayer MoS_2 flakes were grown on sapphire substrates. However, there still exist grain boundaries between two 60° flakes during the CVD process. Due to the different growth conditions, the 60° grain boundaries of chalcogen terminated or metal terminated edges have different structures, which has been displayed in the previous work[78]. **Figure 2.13** illustrates the 60° grain boundaries with chalcogen terminated edges. Very recently, Redwing's group has demonstrated wafer-scale unidirectional WS_2 monolayers on sapphire by MOCVD method[91]. However, they found that there still exists grain boundaries between two unidirectional grains due to the slight tilt of flakes during the growth process. Zhang's group also demonstrated the centimeter-scale single-crystal MoS_2 film on Au foils by CVD approach[92]. The difficulty of fabricating large-scale single-crystal Au foils and the economic cost limit the application of their approach. Thus, more effort need to be devoted to realizing grain boundaries-free TMDCs films by the CVD method.

Chapter 3 : Experimental Techniques

The major experimental techniques for this book will be briefly introduced in this chapter. They are including atomic force microscopy (AFM), scanning tunneling microscopy (STM), and transmission electron microscopy (TEM) for the structural characterization of 2D TMDCs. Also, the optical characterization techniques like Raman spectroscopy, photoluminescence measurement, and second harmonic generation. Here we just discuss the general concept and usage of these techniques for 2D TMDCs. The specific experimental method and setup could be found in the experimental section of each chapter.

3.1 Structural Characterization

3.1.1 Atomic Force Microscopy (AFM)

In brief, Atomic Force Microscopy (AFM) requires a sensitive probe scanning the materials' surface to capture the information and form an image, which is a powerful tool to understand the surface structure of 2D TMDCs materials. The atomic resolution along z-direction clearly displays the steps and thickness of 2D TMDCs, while the high resolution along xy-direction shows a much clear morphology, especially when the 2D crystal with few micrometers or nanometer size. **Figure 3.1 (a)** presents a typical AFM image of 2D TMDCs, the triangle shape of WSe_2 flakes could be recognized due to contrast difference. The height profile in **Figure 3.1 (b)** claims that it is a WSe_2 monolayer structure with a thickness around 0.6 nm. Meanwhile, the homogeneous color of these triangle flakes indicates that they have a similar height.

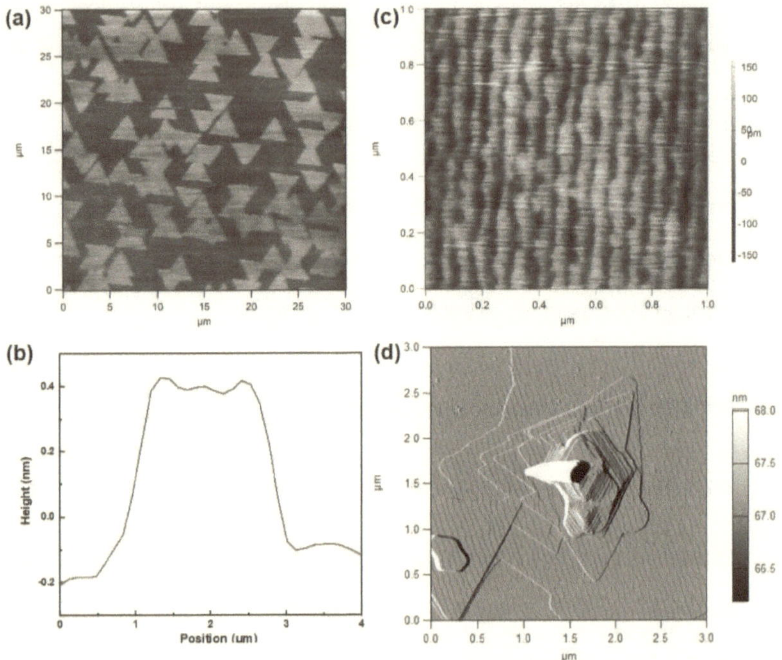

Figure 3.1 Atomic Force Microscopy images of 2D TMDCs. (a) AFM images of monolayer WSe$_2$ triangle flakes. (b) The height profile corresponds to the red line in (a). (c) AFM images of the sapphire surface at a low scale. (d) AFM images of the WSe$_2$ nucleation.

Furthermore, AFM techniques could be applied to characterize substrate surface with picometer steps, which helps to understand the substrate effect during the van der Waal epitaxy process. As shown in Figure 3.1 (c), a sapphire substrate has around 200 pm steps. Except for analysis of the steps, AFM can display the thick nucleation of TMDCs with layered structures or the extremely high (~100 nm) oxides, as shown in Figure 3.1(d). These futures are hard to be recognized by using optical microscopy, and there is no damage happening during the testing process, unlike electron beam microscopy.

3.1.2 Scanning Tunneling Microscopy (STM)

Scanning Tunneling Microscopy (STM) is a technique to measure the surface structure in atomic resolution, thus atoms could be "seen" in STM images. STM has a similar setup as the AFM with a sharp tip. Uniquely, the STM probe receives the tunneling current generated due to the Quantum tunneling effect. Since the quantum tunneling effect happens when two metallic materials are very close to each other, the substrates for the STM measurements must be metallic. For 2D TMDCs, the commonly used substrates for STM are gold and highly oriented pyrolytic graphite (HOPG). The gold substrate is always used for exfoliated TMDCs samples because of its relatively flat surface. On the other hand, the HOPG substrate is applied for the epitaxy growth sample due to its thermal stability and comparable lattice constant to TMDCs materials.

Figure 3.2 exhibits the typical atomic structure of the TMDCs lattice with honeycomb structure. The STM image was captured from CVD-grown monolayer WS_2 on HOPG substrate, and the measured lattice constant is around 3.1Å. More importantly, the atomic defect like vacancies, anti-sites, and grain boundaries become visible by the STM techniques. In chapter 5, we utilized the STM images to count the defect density and to recognize the defect type combined with the scanning tunneling spectroscopy (STS) results.

Figure 3.2 Typical atomic structure of 2D TMDCs on STM image.

3.1.3 Transmission Electron Microscopy (TEM)

Transmission Electron Microscopy (TEM) utilizes the electron beam transmitting through the specimen to obtain an image. For TMDCs characterization, ultrathin TMDCs layers need to be transferred onto the TEM grid with a supported film like glassy carbon or holey carbon. As shown in **Figure 3.3 (a) and (b)**, the TEM image has a significantly higher resolution than the optical image. The tiny seed or hundred nanometer-sized flakes could be seen. Furthermore, the atomic structure of materials could be seen in the scanning TEM (STEM) imaging mode. **Figure**

3.3 (c) and (d) show the typical STEM image of a monolayer tungsten diselenide WSe_2, while the enlarged figure in (d) clearly exhibits the tungsten and selenide atoms. Since the STEM is extremely sensitive to the atoms' atomic number, we could clarify the W and Se atoms from the contrast difference. The brighter the dots are, the havier the elements are. In chapter 4, the STEM images were used to characterize the stacking order of bilayer TMDCs, while we could find that the stacked atoms would also enhance the contrast difference. Furthermore, STEM image is also outstanding for recognizing the defect type and density. The red circle in **Figure 3.3 (d)** displays a "vacancy" in the sulfur position corresponding to the V_{2s} defect.

Both STEM and STM techniques require an ultra-clean surface of materials to characterize the atomic structure. However, the materials always need to be transferred onto the TEM grid, while many impurities like polymer residue would contaminate the samples. This contamination strictly limits the visibility of STEM, as shown in **Figure 3.3 (b) and (c)**, horrible contamination on the surface of thin-film of image (a) and the right side of the image (c) covers the original atomic structure of the specimen. Therefore, a transfer method with a clean process is critical for the TMDCs characterization. In this book, both the PMMA and PDMS assisted transfer methods are applied for TEM characterization. More detailed procedures could be found in each chapter.

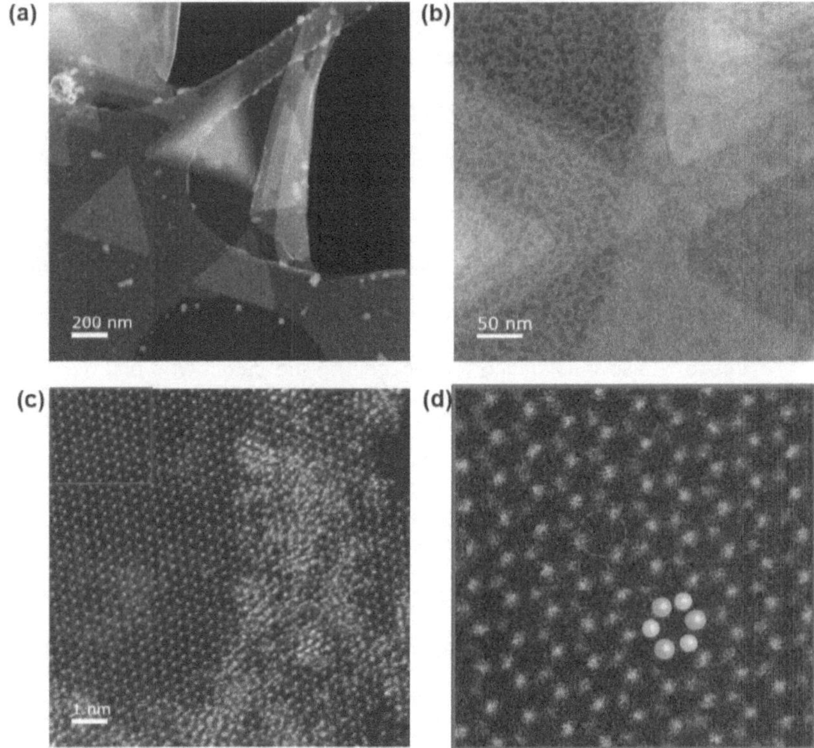

Figure 3.3 TEM images of few-layer TMDCs materials. (a, b) TEM image of CVD-grown WSe_2 transfer onto a holey carbon TEM grid at different scales. (c) STEM image of monolayer WSe_2. (d) Enlarged image from the red squre in (c).

3.2 Optical Characterization

3.2.1 Raman Spectroscopy

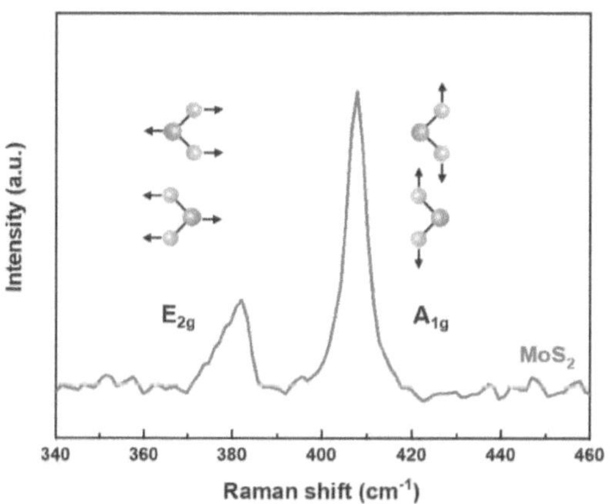

Figure 3.4 Raman spectrum of MoS$_2$ with in-plane and out-of-plane vibration. Raman spectroscopy is a powerful and non-destructive tool for characterizing the structural and electronic properties of 2D materials according to the light scattering effect. First, different materials have different raman peaks, making Raman spectroscopy the optimized technique to quickly identify the TMDCs type without damage. Besides, the raman spectra could provide much other information when carefully analyzing each peak (each peak would correspond to different vibration modes). For example, the in-plane vibration (E$_{2g}$) and out-of-plane vibration (A$_{1g}$) of 2D TMDCs are strongly associated with the defect, dopping, and strain effect on materials (Figure 3.4). Other properties like the layer number, band structures,

interlayer coupling are detectable by Raman spectroscopy. The power and wavelength of the laser source can be tuned, manipulating the raman spectrum and related to the materials properties. In chapter 4, Raman spectroscopy is significant to demonstrate the strain in monolayer WSe_2. Furthermore, the raman position mapping image powerfully shows the heterostructure distribution.

Chapter 4 : Strain-directed Layer-by-layer Epitaxy towards van der Waals WSe$_2$ Homo- and Hetero-structures

Transition metal dichalcogenides (TMDCs) homo- and hetero-stacks hold the tantalizing prospects for integrating as active components for future van der Waals (vdW) electronics and optoelectronics. However, most TMDCs homo- and hetero-stacks are created by onerous mechanical exfoliation, followed by a mixing-and-matching process. While versatile enough for pilot demonstrations, these strategies are not scalable for practical technologies and widespread implementations. Here, we report a two-step epitaxy strategy that promotes second-layer TMDCs on the basal plane of the 1st TMDCs epilayer. The 1st layer TMDCs are grown on substrates where the tensile strength can be tuned by the control of chemical environments. The succeeding epilayers then prefer to grow layer-by-layer on the highly tensile-strained 1st layers. The result is the growth of high-density TMDCs homo-(WSe$_2$) and hetero-bilayers (WSe$_2$-MoS$_2$) with an exceedingly high yield (>99% bilayers) and uniformity. Density function theory (DFT) simulation further sheds light on how strain engineering shifts the subsequent layer growth preference. Second-harmonic generation (SHG) and high-angle annular dark-field scanning transmission electron microscopy (HAADF-STEM) collectively attest to the AB and AA' stacking between the TMDCs epi- and overlayers. The proposed strategy could be a versatile platform for synthesizing diverse arrays of vdW homo- and hetero-stacks, thus providing prospects for realizing the large-scale and layer-controllable 2D electronics.

4.1 Introduction

2D van der Waals (vdW) structures embody a unique class of acritical lattices in which material properties can be manipulated via varying the composition, stacking orders and relative orientation across the atomically thin junctions. Building such artificial vdW junctions is made possible by disassembling bulk crystals into 2D monolayers and reassembling them into artificial bilayer lattices, yielding access to a plethora of exciting physics.[93, 94] The variety of artificial homo- and hetero-structures can be further extended when combined with the recent rediscoveries of transitional metal dichalcogenides (TMDCs), therefore providing a route to previously unachievable semiconductor junctions, superlattices, and light-matter interactions.[95] Notable discoveries include but are not limited to the advent of atomically thin p-n junctions for the ultimate functional unit for nanoscale electronic and optoelectronic devices[96-98], exceedingly high density of states that induce exceptionally high drive currents in the ballistic limit[30, 99], enhanced light absorption with a broader spectral response in UV and NIR regions[100, 101], superior mechanical durability implemented in flexible devices.[102] Despite the immense interest and continuing experimental success by direct mixing and matching monolayer flakes of different materials, widespread implementation of vdW homo- and hetero-stacks has yet to occur. This is primarily due to the difficulty of identifying a general, scalable and reliable epitaxy mechanism that drives the direct growth of vdW homo- and hetero-stacks with control over bilayer coverage and content in tandem with the preservation of crystal qualities.

The general epitaxy growth in chemical vapor deposition (CVD) usually involves the vaporization of precursors, nucleation of TMDCs seeds, and the stitching of individual seeding flakes. In this light, vertical growth of homo- and hetero-structures shall, in principle, be accomplished through the successful growth of a second material on the basal plane of a first layer. To date, significant efforts have been reported to grow few-layered TMDCs in an epitaxial fashion [18, 36, 38, 40, 51, 103]. Ye et.al[18] systematically examined the thermodynamic and kinetic controls of the vertically stacked 2D materials, where the growing conditions, such as annealing temperatures and adatom flux, play crucial roles governing the final structure. Similarly, Li et.al[38] demonstrated that the active clusters with a high diffusion barrier, e.g., the ratio between metal oxides and chalcogenides, modulate the heteroepitaxy growth direction. Besides, a wide variety of growth promoters, including aromatic derivatives and alkali metal halides, has been implemented to regulate the direction of the growth front resulting from the enhanced diffusivity of transition metal oxide precursors. In a nutshell, these researches directed at understanding the mechanisms that underly the epitaxial growth of vdW homo- and hetero-structures predominately hinge on two aspects. One is the thermodynamic control of nucleation on the 1st TMDCs epilayer, and the other is the kinetic modulation of the growth front of the 2nd TMDCs overlayer.[36, 38, 40] However, most discussion is focused on the way to overcome the exceedingly high activation energy on the basal plane of 1st epilayer, little work expounds on what properties of 1st epilayer will result in different transition in growth mode of succeeding TMDCs layer.

Here we report modulating growth modes of succeeding TMDCs layer through creating strain fields of 1st epilayer during the epitaxy growth.[104] The strain level of 1st epilayer was engineered by the dissimilar diffusivity of metal oxide and chalcogen precursors. Comprehensive density functional theory (DFT) calculations indicate that built-in strain fields overshadow the activation energy that predominately dictates the successive growth mode of the epilayer. Atomically resolved images show that the 2nd overlayer grown on top of the strained 1st epilayer has the propensity to follow both AA' (2H) and AB (3R) stacking, respectively. Since the strain on TMDCs can be built on substrates in the CVD growth and is not limited to the current material combinations, the finding here represents an apparent nexus between epitaxial growth and advanced nanoengineering that has the potential to open new opportunity in the scalable production of TMDCs homo- and hetero-stacks.

4.2 Experimental Section

4.2.1 Growth of monolayer, homo- and hetero- bilayer TMDCs

First, the epitaxy growth of single-crystal 1st WSe_2 epilayers was adopted from the previous work[7]. In essence, WO_3 (Sigma-Aldrich, 99.9%) powders were placed in a ceramic boat located at the center of the furnace. The sapphire substrates were located on the downstream side adjacent to the ceramic boat, as shown in Figure 1a. A heating tape annealed the Se powders (Sigma-Aldrich, 99.99%) at the upstream side, and the resulting Se vapors were transported by Ar/hydrogen (H_2) mixture gas (66 to 5 s.c.c.m.). The furnace was then heated to 900°C with a heating rate of 25°C/min at 10 torrs. The duration of epitaxy growth of 1st WSe$_2$ was ~15 minutes, subject to the variation in ramping rate and loading of precursors. Upon completion, the system was naturally cooled down to room temperature. The production of both HTS- and LTS-WSe$_2$ was accomplished by adjusting the precursor ratio between W and Se. For obtaining the WSe$_2$ homo-bilayers, the resultant HTS-WSe$_2$ along with sapphire substrate was directly used as the template and was placed on the downstream side. Note that the growth of 2nd epilayer was carried out under a Se-rich environment. Through controlling the growth time, a fully covered WSe$_2$ homo-bilayer film can be achieved. In the case of synthesizing the MoS$_2$/WSe$_2$ hetero-bilayers, the as-grown HTS-WSe$_2$ flakes/films were transferred into another CVD furnace to avoid cross-contamination. The furnace temperature was set as 800°C with the surface temperature of growth substrate measured around 750~760°C) and the duration of epitaxy growth lasted for 10 minutes. Importantly, in the MoS$_2$ furnace, S powders (Sigma-Aldrich, 99.9%) and MoO$_3$ powders (Sigma-Aldrich, 99.9%) were used as precursors. The flow of carrier gas (Ar) was carefully maintained at 70 sccm, and the pressure was controlled at 30 torr.

4.2.2 Optical characterization of bilayer TMDCs

Raman and photoluminescence (PL) spectra on WSe$_2$ homo-bilayers and WSe$_2$/MoS$_2$ hetero-bilayers were collected using a Witec alpha 300 confocal Raman microscope equipped with a RayShield coupler. A 532-nm solid-state laser as the excitation source. The excitation light with a power of 2.5 mW was focused onto the sample by a 100X objective lens (N.A. = 0.9). The signal was collected by the same objective lens, analyzed by a 0.75-m monochromator, and detected by a liquid-nitrogen-cooled CCD camera. The atomic force microscopy (AFM, Cypher ES-Asylum Research Oxford Instruments) characterizations were conducted with Olympus (OMCLAC240TS) Al-coated silicon cantilevers. The resonance frequency was ~70 kHz, the spring constant was ~2 N/m, and the tip curvature radius was ~7 nm. The temperature-dependent PL spectra on monolayer HTS- and LTS-WSe$_2$ flakes were measured by a micro-photoluminescence setup integrated with a cryogen-free cryostat with a base temperature of 4.2 K. A He-Ne laser with a wavelength of 632.8 nm was focused by a 100X objective lens (NA=0.82) to excite the WSe$_2$ flakes. The PL signals were collected by the same objective lens and sent to a liquid-nitrogen-cooled CCD detector.

SHG spectroscopy was conducted by a back-scattering optical microscope at room temperature[105]. The fundamental field was provided by a mode-locked Ti: sapphire pulsed laser (870 nm) and focused on the sample surface by a 100X objective lens (N.A.= 0.9). The back-scattered SHG signals were collected by the same objective lens, analyzed by a 0.75 m monochromator, and detected by a liquid nitrogen-cooled CCD camera. The polarization of fundamental laser (SHG signal) was selected (analyzed) by combining a linear polarizer and a half-wave plate. For spatial SHG mappings, the sample was mounted on a motorized x-y scanning stage with high repeatability of 0.25 μm.

4.2.3 Scanning transmission electron microscopy (STEM) characterization

The homo- and hetero-bilayer samples for STEM were prepared by wet-transfer methods. Specifically, Polymethyl methacrylate (PMMA) was spin-coated onto the as-grown WSe_2 homo-bilayer samples as a supporting substrate. The PMMA/WSe_2 homo-bilayer composites were thoroughly soaked into HF solution, leaving behind the WSe_2 homo-bilayers. After rinsing with copious amounts of deionized water (DI-H_2O), suspended WSe_2 homo-bilayers were scooped onto TEM grids. To enhance the adhesion between WSe_2 homo-bilayer samples and the underlying TEM grids, additional annealing was carried out at 50°C for 1 hour. Then the residual PMMA was entirely removed by a thorough acetone rinsing. HAADF-STEM imaging was conducted at 80 kV using a JEOL ARM 200F transmission electron microscope (80-200 kV). TEM equipped with Cs (spherical aberration) corrector, a high brightness cold field-emission gun (C-FEG), and a STEM detector is made of yttrium aluminum perovskite (YAP).

4.2.4 Density function theory (DFT) simulation

The first-principles calculations were carried out with DFT as implemented in the Vienna Ab initio Simulation Package (VASP).[106] The interaction between electrons and ionic cores was approximated by the projector augmented wave method, and the exchange-correlation potential was described by the PBE-GGA,[107] with the vdW correction vdW-DF (optB86) functionals.[108] A slab model is used for simulation with a vacuum thickness larger than 18 Å to eliminate the spurious interaction, and the plane waves energy cutoff is 400 eV. The structure was fully relaxed until the energy and force change reached 10^{-5} eV per 1 × 1 cell and 10^{-3}

eV/Å, respectively. A $2 \times 2 \times 1$ and $3 \times 3 \times 1$ k-grid was used for low strained ($5 \times 5 WSe_2$ / $2\sqrt{3} \times 2\sqrt{3}$ Al_2O_3) and high strained ($\sqrt{13} \times \sqrt{13} WSe_2$ / $\sqrt{7} \times \sqrt{7}$ Al_2O_3) systems, with an interlayer distance 1.77 Å and 1.70 Å between WSe_2 and Al_2O_3, and interlayer distance between 6.58 Å and 6.5 Å WSe_2 and WSe_2. The lattice constants of WSe_2 and sapphire used for calculations are 3.28 Å and 4.73 Å.

4.3 Results and discussion

4.3.1 Strain engineering for single layer WSe_2

The process for activating the basal plane as the growth front begins with the epitaxial growth of WSe_2 monolayers by CVD that features the dual-heating zones[7] as schematically illustrated in **Figure 4.1a**. Here, the WO_3 source was located at the center of the heating zone 2, and sapphire substrates were placed adjacent to the WO_3 boat. The temperature of the heating zone 2 was fixed at 900°C. The annealing temperature of the heating zone 1, where the Se powders are located, varied from 220°C to 300°C. We found that this Se annealing temperature is strongly associated with the shift in photoluminescence (PL) of the as-grown WSe_2 monolayer specimens that is characteristic of the structural strain (**Figure A1**)[71, 74, 109, 110]. When the Se annealing temperature was set at 220°C, as-grown WSe_2 monolayer shows an average PL emission at around 1.56eV along the whole deposition distance. Conversely, the PL energy shifts to 1.63eV when the Se annealing temperature sets at 300°C. To this end, we set the annealing temperature of first hot zone at a medium temperature of 270°C, which in turn enables us to explore the origin of built-in strain fields. PL spectra were taken from

WSe$_2$ specimens epitaxially grown on sapphire substrates located between the upstream to the downstream regions as shown in **Figure 4.1b**. The distance between each sapphire substrate is evenly separated by 1 centimeter to ensure the comparison of PL spectra and the associated built-in strain fields on a fair footing. To better understand the correlative strain effect, we selected two typical samples at the opposite ends of depositing position profiles as shown in **Figure 4.1c**. Relative to the prototypical emission ~1.66 eV of unstrained monolayer WSe$_2$ reported in literature[74], PL spectra of WSe$_2$ collected near the beginning of the diffusion pathway marked as 0 centimeter displayed a substantial red shift of 74 meV which can be translated into a tensile strain of ~1.2%, designated as high tensile strained (HTS) WSe$_2$.[69, 70] Meanwhile, only a minor red shift of 27 meV was observed from WSe$_2$ specimens grown near the end of the diffusion pathway where Se vapors become predominant. The minor shift accounts for a relatively low tensile strain around 0.43%, namely low tensile strained (LTS) WSe$_2$.[69, 70] Intriguingly, it is found that the tensile strains in both samples relaxed after the wet transfer to freshly cleaned sapphires and their emission peaks revert back to ~1.66 eV. The tensile strain is likely built upon the cooling after growth and it is proposed that the different strains might be related to the thermal coefficient difference between the sapphire substrate and the TMDCs grown with various conditions, which will be further discussed in the following context.

It is known that defects, trions and band-to-band emissions all exhibit characteristic emission evolutions when excited at low temperatures.[111-113] Thus, to rule out the defect- and doping-induced shift in PL emission, we performed low-temperature

PL measurements as shown in **Figure A2**. As suggested in the Gaussian fitted spectra, we did not observe distinctive PL emission difference stemmed from defect and doping contributions of both HTS- and LTS-WSe$_2$ samples. These results further attest to the fact that peak position and intensity alteration are more likely caused by strain-induced band variation. (Detailed discussion in **Supplementary Discussion 1**) Furthermore, based on temperature-dependent photoluminescence measurement (**Figure A3 and A4**), we noted that the PL energy of both samples all displayed a large redshift with the increasing temperature, as a result of the increased electron-phonon interactions and varied bonding lengths. The HTS sample can be considered that the WSe$_2$-sapphire exhibits a stronger interaction, while the low tensile strained sample LTS exhibit a much weaker (van der Waals-like) interaction. The mismatch in the thermal expansion coefficient between 1st epilayer (WSe$_2$, 1.1×10^{-5}/K-1.4×10^{-5}/K) and underlying substrate (sapphire, 5×10^{-6}/K-8.3×10^{-6}/K) has profound impacts in the HTS-WSe$_2$ when the temperature of PL measurement was decreased from 300K to 4K[114]. The result is the much wider PL energy difference between the HTS and LTS samples increased from 40 meV (300K) to ~60 meV (4K). To this end, we employed a modified Varshni relationship to quantitatively investigate the temperature dependence on the bandgap of strained WSe$_2$[115, 116]:

$$E_g(T) = E_g(0) - S\langle\hbar\omega\rangle\left[\coth\left(\frac{\langle\hbar\omega\rangle}{2k_BT}\right) - 1\right] \quad (1)$$

, where E_g (0) is the bandgap energy at absolute zero temperature, S is a dimensionless constant that represents the strength of the exciton-phonon coupling, $\langle\hbar\omega\rangle$ describes the average phonon energy involved in electron-phonon interactions, and $\hbar$ is Plancks constant and k_B is the Bolzmann constant. For HTS-WSe_2, the fitting parameter was extracted with E_g (0) ~1.67 eV, S ~2.07, and $\langle\hbar\omega\rangle$ ~11.71 meV. Similar fitting to LTS-WSe_2 yields E_g (0) ~1.73 eV, S ~2.66, and $\langle\hbar\omega\rangle$ ~18.80 meV. By comparing these parameters, the HTS sample was found to exhibit a lower bandgap value and weaker exciton-phonon coupling, agreeing well with the previously reported results where monolayer WSe_2 is subjected to tensile strain.[117] Meanwhile, Raman spectroscopy was used to study the spatial distribution of the built-in strain of the as-grown WSe_2. **Figure 4.1d** demonstrates that the E_{2g} peak of WSe_2 splits into two peaks and shifts to opposite directions when the growth of WSe_2 carries out near the upstream. This is characteristic of the emergence of relatively higher tensile strain[110] which is also typically seen in CVD-grown single crystal TMDCs[118, 119]. Meanwhile, when the growth of WSe_2 takes place at the downstream region, the splitting of the doublet peak gradually narrows and eventually merges at ~250 cm^{-1}, indicative of a nearly strain-free state.[110] These results collectively attest to the fact that peak position and intensity alteration are more likely caused by strain-induced band variation during the epitaxy.

As mentioned in the previous section, the root cause responsible for the strain variation is the combined results of subtle differences in growth environments. It is known that the diffusion of metal (oxide) vapors in conjunction with a significant vapor concentration drop always gives rise to a non-uniform growth distribution when the distance between metal oxide sources and designated substrates increases[34]. On the other hand, Se exhibits a low melting point (220°C) and thus can diffuse throughout the whole furnace tube when the furnace temperature reaches 900°C at 10 torr.[34] We performed the evaporation of WO_3 powders in a standard growth condition except for the absence of Se (see **Figure A5** for SEM images). Energy-dispersive X-ray analysis (EDX) is used to examine the relative W weight percentage of the deposited film along the path of diffusion. **Figure 4.1e** clearly shows that the concentration profile of W distributes in a declining fashion, suggesting that the WO_3/Se vapor ratio scales disproportionally with the increasing distance between growth location and WO_3 source. This concentration profile shows a similar trend compare to the energy distribution shown in Figure 1b. In the meantime, we also performed the 1st WSe_2 epilayer growth on top of SiO_2/Si substrates as described in **Figure A6** and **Supplementary Discussion 2**. On the basis of the experimental observations and spectroscopic characterizations, we have arrived at the conclusion that the WO_3/Se vapor concentration profile enables the modulation of the strain levels in monolayer WSe_2. Under the Se-deficient atmosphere (growth takes place near WO_3 boat or at low Se heating temperature), the reduced W atoms may not be fully selenized and then completely converted into WSe_2. Hence, fractions of W atoms in the WSe_2 flake are chemically tethered with the oxygen-terminated sapphire surfaces at the reaction temperature. Next, the chemically pinned WSe_2 tends to develop the high built-in strain by virtue of the mismatch of thermal expansion coefficient between that of WSe_2 and sapphire when cooling down from a high annealing temperature in the growing process.[114] At the same time, the Se-rich environment efficiently transforms WO_3 precursors to WSe$_2$, where the unsaturated W-O bond formation enables WSe$_2$ monolayers to find an energetically favorable configuration and thus leads to a low tensile strained monolayer. We note that the deposition temperature near the upstream location is determined to be 925°C, and it gradually decreases to 892°C when the depositing position moves to the downstream site. (**Table A1**). As discussed in **Supplementary Discussion 3**, the temperature difference measured at 0 cm and 9 cm would be nuance (~0.03%) and less likely to contribute to the built-in strain on the 1st WSe$_2$ epilayer.

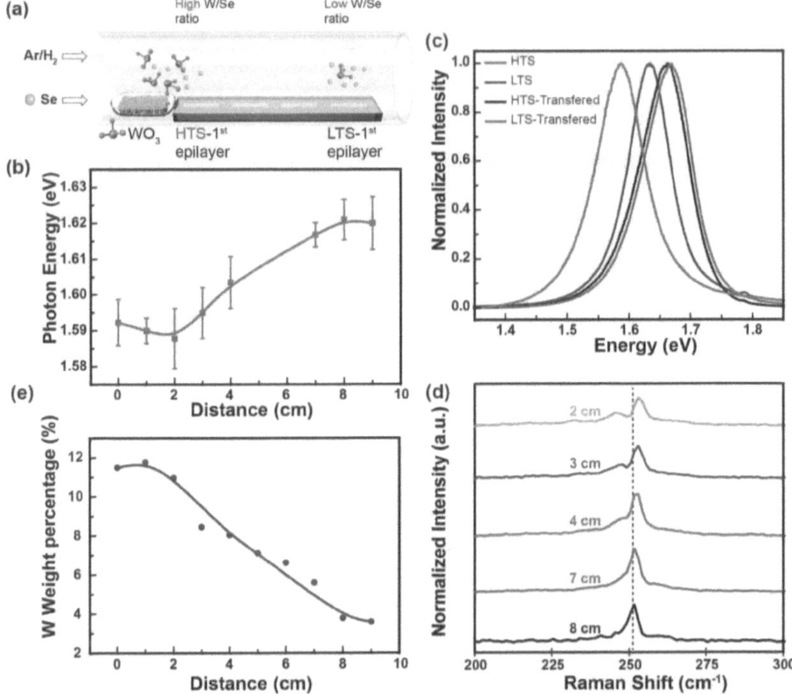

Figure 4.1 Strain engineering of 1st WSe$_2$ epilayers. (a) Schematic illustration shows a representative CVD epitaxy growth of 1st WSe$_2$ epilayers. (b) Photon Energy distribution of 1st WSe$_2$ epilayers collected at different distances away from the WO$_3$ source. Error bars are included to provide a standard deviation of photon energy. Data were collected from over fifty samples. (c) Photoluminescence (PL) spectra of high tensile strained (HTS) and low tensile strained (LTS) 1st WSe$_2$ epilayers grown on sapphire substrates before and after transferring at room temperatures. (d) Raman spectra of the as grown 1st WSe$_2$ epilayers collected at different distances from away WO$_3$ source. (e) The concentration distribution of W (blue) as a function of different distances away from WO$_3$ source. The weight percentage of W with respect to detected elements (Al, O, and W) based on EDX data.

4.3.2 Strain-induced epitaxy of WSe$_2$ homo-bilayers

To elucidate how tensile strain modulates the homoepitaxial growth of WSe$_2$ bilayer, we applied a two-step growth procedure where monolayer WSe$_2$ is grown on sapphire followed by the van der Waals epitaxy normal to the basal plane growth front. The ability to systematically manipulate the built-in strain fields enables us to prepare the WSe$_2$ monolayers with various tensile strain levels as the templates to direct the growth of the subsequent 2nd WSe$_2$ overlayer. As summarized in **Figure 4.2a**, the built-in strain fields of the 1st epilayer WSe$_2$ indeed plays an important role in modulating the epitaxial growth front for the formation of WSe$_2$ homo-bilayers. The HTS-WSe$_2$ monolayer enables the layer-by-layer growth mode, whereas the LTS-WSe$_2$ directs the island growth mode. **Figures 4.2b** and **4.2e** feature the optical microscope (OM) images of the 1st epilayer of HTS- and LTS-WSe$_2$, respectively. PL and Raman measurements were used to confirm the strain uniformity and were found to follow the similar trend (see Supporting Information **Figure A7**). Here, the 1st HTS-WSe$_2$ monolayer template, even with the built-in strain of only 1.37%, evidentially and sufficiently enables the layer-by-layer growth mode, whereas the epitaxy on LTS-WSe$_2$ (a built-in strain of 0.56%) favors the island growth mode.

After the 2nd growth, a uniform and fully covered 2nd overlayer of WSe$_2$ epitaxially grown on top of the HTS-WSe$_2$ epilayer as highlighted by the dashed triangle in red as shown in **Figure 4.2c**. AFM image in **Figure 4.2d** provides the close-up view that the growth proceeds with the seeding of small triangular WSe$_2$ and then merges into a complete WSe$_2$ bilayer. The height profile determined by a blue dashed line across the interface between substrate and the 1st epilayer of WSe$_2$ flakes is of 0.82 nm, and the second layer is around 0.62 nm, in accord with the signature thickness of monolayer WSe$_2$.[7] Statistically, after scouring over 1,000 flakes from various epitaxy batches, all of the HTS-WSe$_2$ templates proceed with the layer-by-layer growth mode without exception. **Figure A8** showcases the low magnification OM image where WSe$_2$ homo-bilayers with uniform coverage are ubiquitous on all HTS-WSe$_2$ epilayers. To the contrary, sporadically distributed islands or multilayer clusters (thickness up to ~10 nm) were found to scatter all over the LTS-WSe$_2$ flakes as shown in **Figures 4.2f** and **4.2g**. Note that although the activation of basal plane prioritizes the vertical growth front, the underlying HTS-WSe$_2$ templates concurrently become discernably larger. The dashed triangles in Figure 2c and 2f presents the size of 1st epilayer in Figure 2b and 2e. Here, the edges of the first layer WSe$_2$ serves as the nucleation sites for the 2nd in-plane growth of WSe$_2$ during the ramping process.[57, 120] To ensure the yield and coverage of the 2nd WSe$_2$ overlayer, epitaxy growth should be conducted in the high-temperature region (~900°C) where adequate energy overcomes the nucleation barrier on the basal plane of the HTS-WSe$_2$ templates, therefore proceeding with the thermodynamically favorable stacking structure afterward.[18, 36] Epitaxial growth at the lower temperature leads to edge-directed in-plane growth (See **Figure A9**) and ultimately extends to the continuous WSe$_2$ thin films. **Figure A10** shows a typical distribution of second-growth results of HTS-WSe$_2$ epilayers. At the very front end, where the HTS-WSe$_2$ sample is placed right next to the W source, continuous multilayers (3-4 layers) were developed on account of the sample precursor supply. The number of layers sequentially decreases to 2-3 layers and ultimately limits to only bilayers when the HTS-WSe$_2$ sample gradually moves away from W source. The same trend was also observed in the case of low-pressure CVD monolayer growth.[121] It is noted that the scattered bright spots are thick nucleation sites. The presence of thick seeds can be suppressed efficiently by increasing Se vapor flux or decreasing the H$_2$/Ar ratio to restrain the W supply during the 2nd growth.

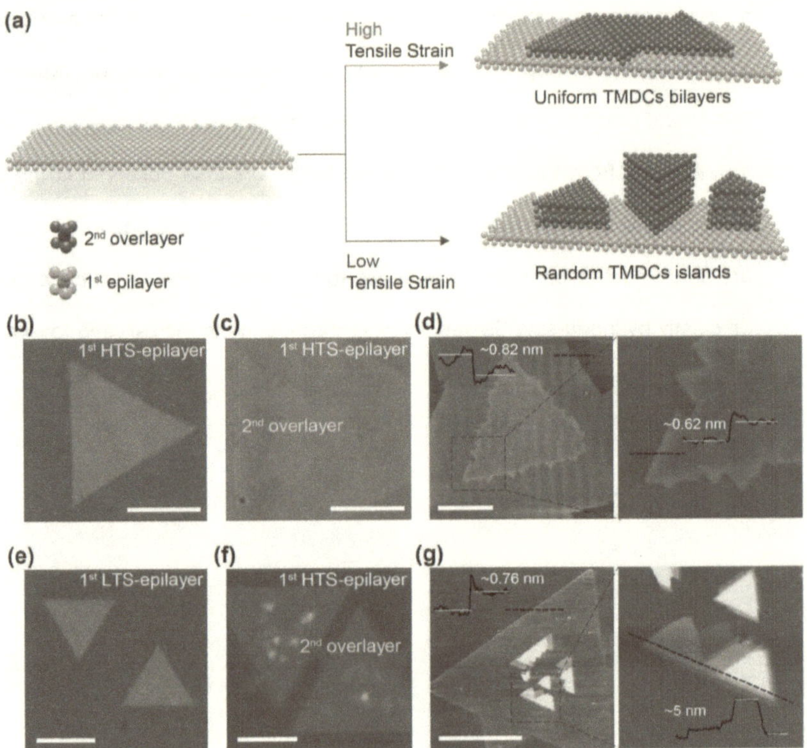

Figure 4.2 Growth mode control by tensile strain. (a) Schematic representation features the growth of homogeneous 2nd WSe$_2$ overlayers on the 1st HTS-WSe$_2$ epilayer in a layer-by-layer fashion. Meanwhile, island growth of multilayered clusters was found on the basal plane of the LTS-WSe$_2$. (b) Optical microscope (OM) image of the as-grown (0cm) HTS-WSe$_2$ monolayer. (c) OM showing the formation of WSe$_2$ homo-bilayer indicated by the red, dashed triangle. (d) Corresponding atomic force microscope (AFM) images of WSe$_2$ homo-bilayers as well as a close-up view of the blue-marked area. Inset shows the height profile along the black dotted line. (e) OM image of the as-grown (9cm) LTS-WSe$_2$ monolayer. (f) OM of WSe$_2$ islands formed during the 2nd growth. (g) Corresponding AFM images reveal the multilayered islands scatted on the basal plane of 1st LTS-WSe$_2$ epilayers. Scale bars: 5 µm.

4.3.3 Characterization of multilayer WSe_2

The successful activation of the basal plane through strain engineering of 1st WSe_2 epilayer guides the layer-by-layer growth to obtain bilayer and even the trilayer as shown in **Figure A11**. AFM images reveal that the triangular WSe_2 seeds first nucleate, merge with the adjacent flakes, and ultimately form a complete film prior to the growth of subsequent layers, closely resembling the layer-by-layer growth mode. Aside from the consistently aligned vertical growth on basal plane, the 1st HTS-WSe_2 templates and 2nd WSe_2 overlayers are highly aligned and the dominant the twist angles/edge orientations are 0° and 60° (180°). To verify the orientation of individual flakes, the WSe_2 homo-bilayers were characterized by second-harmonic generation (SHG). It is known that polarization resolved SHG is very sensitive to the crystal orientation and the intensity profile map can be used as a descriptor for verifying spatial orientations between 1st HTS- WSe_2 epilayer and 2nd WSe_2 overlayer. With this in mind, we specifically selected WSe_2 homo-bilayers with incomplete coverage of 2nd WSe_2 overlayers as shown in **Figure A11a** to further enhance the fidelity of grain identification. **Figure 4.3a and 4.3b** show the OM image and the corresponding SHG intensity map of the triangular 2nd WSe_2 overlayers, which is independent of the polarization direction of the incident laser field with respect to the crystallographic axis. Here, the total SHG intensity of the sample was recorded. It is widely recognized that materials with the absence of inversion symmetry give rise to the SHG signal. Here, monolayer WSe_2 is characterized by the D_{3h} point group symmetry which enables a strong SHG signal by virtue of its broken inversion symmetry. As a result, in the case of WSe_2 homo-bilayers, the enhanced SHG intensity plot (bright) stemmed from a parallel stacking (0°) is the result of 3R-stacking or AB stacking which can be assigned to the D_{3h} point group with broken inversion symmetry. On the contrary, the much suppressed second harmonic intensity plot (dark) taken at an anti-parallel stacking (60°) is the consequence of the Bernal (2H) stacking or AA' stacking (the D_{3d} point group) with inversion symmetry.[105] We further examined the polarization-resolved SHG as shown in **Figure 4.3c**. The polarization direction for bilayer homostructures with different twist angles seems to align well with the 1st HTS-WSe_2 templates, an indication of the formation of energetically and thermodynamically favorable stacking between the 2nd overlayers on the 1st HTS-WSe_2 (0° or 60°). In addition, we extended the SHG measurements to different batches of WSe_2 homo-bilayers. From analyzing the compiled statistics shown in **Figure 4.3d,** it becomes clear that the stacking mode only shows two directions (0° twist or 60° twist) that are generally recognized as the most stable stacking in WSe_2 system.[40, 43] These results are consistent with the previous observation on artificially stacked bilayer MoS_2.[105]

In parallel, the stacking configurations of the WSe_2 homo-bilayers were investigated by the plane-view scanning transmission electron microscopy (STEM) with the high-angle annular dark-field (HAADF) mode. This is due to the fact that Z contrast is highly dependent on the atomic numbers, making it possible to infer the detailed atomic arrangement. Here, WSe_2 homo-bilayers comprised of opposite twist angle (0° and 60°) were selected (**Figure A12a**). **Figure 4.3e** and **4.3g** feature the high-resolution HAADF-STEM images taken at WSe_2 homo-bilayers of 0° and 60° twist angles. Distinctively different atomic patterns become discernible as a result of different stacking sequences as shown in **Figure A12b** and **A12c**. In the AB stacking configuration (twist angle of 0°, **Figure 4.3i**), atoms intersected by the red dash-line represent the 2Se+W overlaps. The bright spots marked in **Figure 4.3e** manifest the overlaps by means of sandwiching W atoms with 2 Se atoms in the vertical and aligned fashion. Furthermore, from the sectional view of the WSe_2 bilayer stacking, it becomes apparent that the upper layer slides and thus gives rise to three atomic alignments, including 2Se,W, and 2Se+W (**Figure 4.3j**). The intensity profile extracted from the selected regions in **Figure 4.3e** also follows the similar trend in terms of stacking sequence as shown in **Figure 4.3f**. Alternatively, atomic configuration featured in **Figure 4.3k** with a 60° rotation of the hexagonal structure results in the alignment of W with Se, i.e., the AA' stacking with the atomic alignment of W+2Se and 2Se+W (**Figure 4.3l**), respectively. The intensity profile (**Figure 4.3h**) derived from **Figure 4.3g** matches well with the aforementioned atomic stacking sequence.[44] Moreover, the optical properties of as-grown WSe_2 with different layer number are systematically investigated (**Figure A13**). The results suggest that the growth front through activating the basal plane during the epitaxy has made possible the preservation of all these characteristics unique to the exfoliated benchmarks, thus distinguishing the strain-engineering approach from the other strategies.[38]

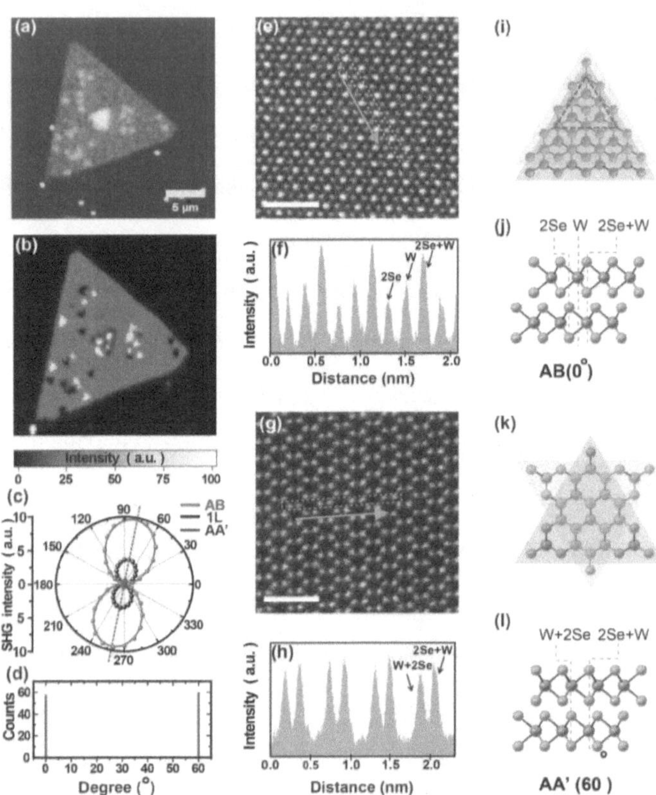

Figure 4.3 Stacking configuration of WSe₂ homo- bilayers. (a) The optical microscope (OM) image shows the inception of triangular 2nd WSe₂ overlayers grown on 1st HTS-WSe₂ epilayer. (b) The corresponding false-colored second harmonic intensity obtained by pixel-to-pixel spatial mappings in (a). (c) Polar plots of the polarization-resolved second harmonic intensity as a function of azimuthal angle φ measured from as-grown sample. (d) Statistical histograms of the twist angles measured from hundreds of the triangular 2nd WSe₂ overlayers. (e) HAADF-STEM image of WSe₂ homo- bilayers with a 0° twist angle. (f) Intensity profile along the direction of selected line in (e). (g) HAADF-STEM image of WSe₂ homo-bilayers with a 60° twist angle. (h) Intensity profile along the direction of selected line in (g). Atomic configuration of AB (0° twisted, 3R) stacked WSe₂ homo-bilayer, top- (i) and cross-sectional views (j). Atomic configuration of AA' (60° twisted, 2H) stacked WSe₂ homo-bilayers, view from the top (k) and the side (l). Blue balls denote the W atoms and yellow balls represent the Se atoms in i, j, k and l. Scale bar: 1 nm.

4.3.4 Mechanism of strain-induced bilayer WSe$_2$

To understand the correlation between strain and the associated growth modes, DFT computations were performed to probe the information of formation energy in the presence of strain. To this end, atomic models of monolayer WSe$_2$ epitaxially grown on sapphire with various degrees of the magnitude of built-in strains ranging from -1% to 6% were constructed. The formation energy is calculated for each superstructure that sits at the lowest energy among various atomic registries by sliding the layer surface with respect to the substrate. The calculated formation energy on the basal plane, $E_f = E_{WSe_2/Al_2O_3} - E_{WSe_2} - E_{Al_2O_3}$, is found to decrease linearly as strain increases (**Figure 4.4a**), suggesting that the high tensile strain indeed arises from the strong interaction between WSe$_2$ and sapphire substrate. The formation energy E_f is known to be associated with the change of Gibbs free energy per unit volume of the solid phase, ΔG_v. The change of volume of Gibbs free energy can be expressed as, $\Delta G^* = \left(\frac{16\pi\gamma Vf}{3(\Delta Gv+w)^2}\right) \times \left(\frac{2-3cos\theta+cos^3\theta}{4}\right)$, where γ is the surface tension, θ represents the contact angle, and w denotes the strain energy per unit volume generated by the stress in the epitaxy. When the strain energy per unit volume generated by the disproportional concentration profile of WO$_3$/Se precursor vapors is included, the overall energy barrier to nucleation increases because the sign of ΔG_v is negative and the sign of strain energy is positive.

With this in mind, as shown in **Figure 4.4b**, we investigate the energetics of layer-by-layer growth mode of the subsequent 2nd WSe$_2$ overlayer. Two atomic models for the LTS (5×5WSe$_2$ / $2\sqrt{3}\times2\sqrt{3}$ Al$_2$O$_3$) and HTS ($\sqrt{13}\times\sqrt{13}$ WSe$_2$ / $\sqrt{7}\times\sqrt{7}$ Al$_2$O$_3$) WSe$_2$ on sapphire corresponds to the two extremities of the WSe$_2$-sapphire interactions: (i) nearly strain-free (-0.1%), and (ii) extremely high orders of magnitude of strain level of 5.8%, respectively. **Figure 4.4c** shows the calculated energy cost of the formation of nth (n=1,2,3) WSe$_2$ overlayer (with regard to free-standing single layer) on the two aforementioned systems, defined by $E_n = E_{(n)WSe_2/Al_2O_3} - E_{(n-1)WSe_2/Al_2O_3} - E_{free-WSe_2}$. Another two reference cases, HTS- and LTS-WSe$_2$ in a freestanding form, i.e., without the supporting Al$_2$O$_3$ substrates, are also included for comparison. It is found that the energetic landscape of the subsequent epilayer always shifts toward the lower end in the less strained system (the difference is around 260 - 270 meV per chemical formula for n=2 and 3). Meanwhile, after including the underlying Al$_2$O$_3$ growth substrates, the energy cost of the 1st WSe$_2$ epilayer for the high tensile strain case becomes even lower than that of low strain case, shifting by 112 meV per chemical formula. This shift in energy cost thus suggests the presence of strong interaction between 1st WSe$_2$ epilayer and Al$_2$O$_3$ in the high tensile strain cases. (Note that the distance between bottom Se layer and O-termination surface is 1.77 Å and 1.7 Å for low and high strain systems). The energy cost for adding 2nd and 3rd WSe$_2$ overlayers becomes nearly identical with or without substrates (**Figure 4.4c**). In parallel, contributions to the layer-by-layer energy cost of WSe$_2$ /Al$_2$O$_3$ system are two-fold: substrate interaction and strain effect, with the former being negative, and the latter being positive. The stronger interaction with underlying substrate in HTS-WSe$_2$ induced Frank–van der Merwe (layer-by-layer) growth mode[122], adatoms attach preferentially to surface sites resulting in lateral grown film prior to growth of subsequent layers along vertical direction. Antithetically, a negligible strain in LTS-WSe_2 leads to Volmer–Weber (nucleation-to-island) growth[123], indicating very weak interaction with surface, such that adatom–adatom interactions are stronger than those of the adatom with the surface, resulting in the vertical growth with adatom clusters or islands. As a result, the layer-by-layer vertical growth of WSe_2 bilayers is energetically preferred at the beginning of nucleation stage and becomes thermodynamically stable at the end of epitaxial growth by virtue of compensating the energy cost induced by the high tensile strain ($E_{2,3}>0$). For low strain (nearly strain-free) systems, the total energy cost is always negative and therefore the growth of subsequent 2nd WSe_2 overlayer favors the nucleation-to-island mode. Our experimental observation and simulations agree well with above thin film growth models.

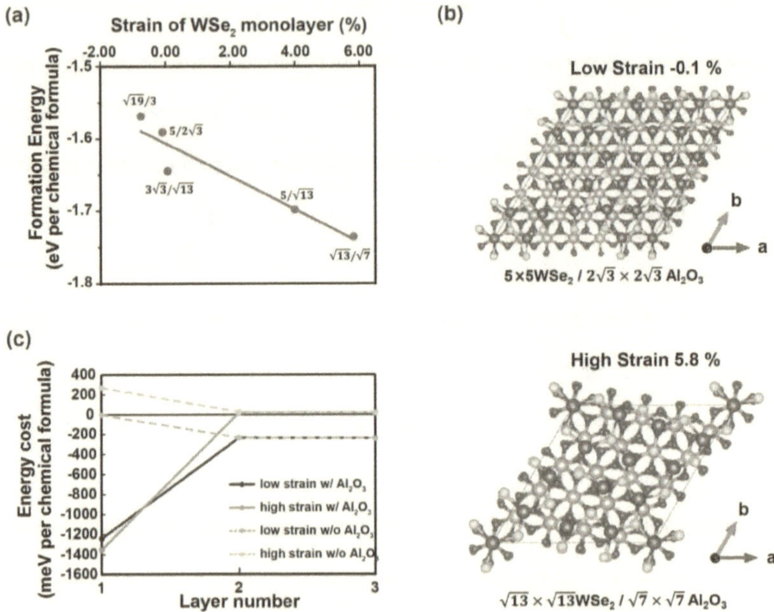

Figure 4.4 Simulation toward understand the growth mechanism. (a) Formation energy of monolayer WSe2 on sapphire (Al_2O_3) superstructures with various strain. (b) Atomic models for the low strained ($5 \times 5 WSe_2 / 2\sqrt{3} \times 2\sqrt{3} Al_2O_3$) and high strained ($\sqrt{13} \times \sqrt{13} WSe_2 / \sqrt{7} \times \sqrt{7} Al_2O_3$) WSe2 on Al_2O_3. (c) The calculated energy cost of nth layer WSe2 (with respect to the free standing monolayer WSe2) on two systems with substrates, defined by $E_n = E_{(n)WSe_2/Al_2O_3} - E_{(n-1)WSe_2/Al_2O_3} - E_{free-WSe_2}$ and other two without substrates.

4.3.5 Growth and characterization of MoS2/WSe2 hetero-bilayers

Because our strain-engineering epitaxy is directed by the built-in strain fields, intrinsic to the 2D TMDC growth, its use is not limited to the WSe2-WSe2 homo-bilayers reported here. Instead, it could be generalized for epitaxy growth of vdW hetero-bilayers with atomically thin p-n junctions. Here, combined with the understanding of the nucleation and growth characteristics on HTS-WSe$_2$ template, we expanded the stain-induced epitaxy growth to include MoS$_2$/WSe$_2$ p-n hetero-bilayers. To this end, HTS-WSe$_2$ 1st epilayers were again featured as the strained template for the subsequent growth of 2nd MoS$_2$ overlayers. Note that a two-step growth in different furnaces is used to afford the heterostructures without cross contamination at the atomically thin p-n interfaces. The growth temperature of 2nd MoS$_2$ overlayers is set at 750~760°C under 30 torr with Ar as the carrier gas. **Figure 4.5a** schematically illustrates the sequential epitaxy growth of 2nd MoS$_2$ overlayers. Both vertical and lateral epitaxy growth of MoS$_2$ on the HTS-WSe$_2$ templates are observed with the vertical growth front (basal plane) outweighed the lateral counterparts (edges). This can be understood from the epitaxial point of view that the formation of lateral p-n junctions takes place during the ramping of annealing temperatures for the 2nd growth.[57, 124] We specifically selected WSe$_2$/MoS$_2$ hetero-bilayers with an incomplete coverage of 2nd MoS$_2$ epilayer as a result of PL quenching which can be leveraged as a descriptor to identify the location of 2nd MoS$_2$ epilayers. **Figure 4.5b** shows the OM image of the as-grown MoS$_2$/WSe$_2$ hetero-bilayers. The distribution of the 1st WSe$_2$ template and the 2nd MoS$_2$ overlayer was further examined by PL intensity maps at 1.59 (wavelength of 780 nm) and 1.88 eV (wavelength of 660 nm), characteristic of the direct excitonic transitions in both MoS$_2$ and WSe$_2$. PL mapping of relevant PL characteristics, including MoS$_2$ in green (**Figure 4.5c**) and WSe$_2$ in red (**Figure 4.5d**), evidentially proves the successful out-of-plane growth of n-type MoS$_2$ on the basal plane of p-type WSe$_2$. In addition, the PL spectra in **Figure 4.5e** reveals diverse PL emission peaks at the core (blue) and from the surrounding (black) regions. The black curve with a prominent emission peak at 1.88 eV represents the direct bandgap excitonic peak of monolayer MoS_2. Meanwhile, the emergence of two relatively weak peaks in blue (1.59 eV vs. 1.85eV) was observed in the core region. The peak assigned at 1.59 eV corresponds to the combination of intralayer excitonic emission from the underlying WSe_2 epilayer (X_{WSe2} at 1.61eV) and the MoS_2/WSe_2 interlayer exiton ($X_{MoS2/WSe2}$ at 1.58 eV), as shown in **Figure A14**.[125] The considerable deviation and intensity quenching of PL emission can be translated into the strong interlayer coupling and charge transfer, characteristic of the type II heterostructures.[126] Similarly, Raman spectra acquired from the core region (black) and surrounding (blue) area of the MoS_2/WSe_2 heterostructures were summarized in **Figure 4.5f.** Two predominant peaks at 405 and 385 cm^{-1}, which correspond to the A_{1g} and E_{2g} modes of monolayer MoS_2 arise from the surrounding. Nevertheless, the core region exhibits both characteristic peaks of MoS_2 and WSe_2 (250 cm^{-1}, E_{2g} mode), respectively. The presence of relevant Raman characteristics in tandem with representative features from both PL spectra and mappings collectively manifests the applicability and generality of strain-induced epitaxy growth of hetero-bilayers with atomically thin and clean p-n interfaces. A wide variety of artificial TMDCs hetero-stackings with control over number of layers, uniformity, coverage and intrinsic material properties can thus be epitaxially grown for a myriad of electronic and optoelectronic applications.

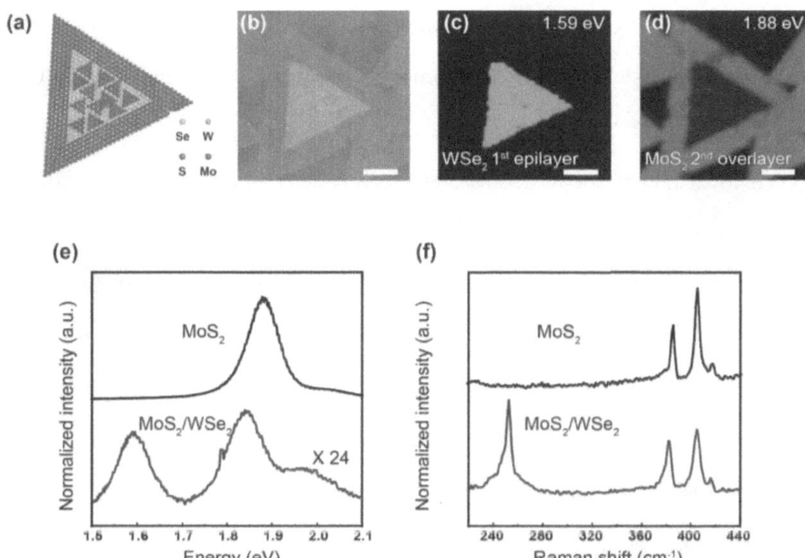

Figure 4.5 MoS₂/WSe₂ heterostructure obtained from 1ˢᵗ HTS-WSe₂. (a) Schematic illustration of atomic structures and (b) optical microscope (OM) image of vertical and lateral MoS₂/WSe₂ heterostructures. (c, d) The corresponding PL mapping image at 1.59 eV and 1.88 eV, respectively. (e) PL spectra and (f) Raman spectra corrected from the core (blue) and surrounding (black) regions. Scale bars: 5 µm

4.4 Conclusions

In conclusion, we have demonstrated that built-in strain fields in 2D TMDCs play a crucial role in activating the out-of-plane growth front for affording homo- and hetero-bilayers. With the consistent agreement between DFT simulations, experimental observations, and spectroscopic characterizations, our findings elucidate general rules for strain-driven epitaxy growth of homo- and hetero-bilayers: (1) the overall energy barrier to rampant nucleation increases because of the presence of strain energy that reconciles the adverse impact of ΔG_v. The result is the activation of the growth front on the basal plane where epitaxy of bilayer proceeds, (2) built-in strain fields in the 1st WSe$_2$ epilayers stem from the combined effects of local chemical environments and can be systematically modulated in situ during the epitaxy growth, (3) the stacking of homo- and hetero-epilayers predominately follows AB (0° twisted, 3R) and AA' (60° twisted, 2H) according to the analyses of SHG measurement and HAADF-STEM images, and (4) this strain-modulated epitaxy can be a general platform for extending libraries of vdW homo- and hetero-bilayers. Such an understanding not only provides fundamental insights into the knowledge for vapor-phase epitaxy of 2D layers but opens new inroads in the direct growth of vdW homo- and hetero-bilayers with control over bilayer coverage and content in tandem with the preservation of electrically addressable and spectroscopically tractable interfaces in a potentially scalable, and industry-compatible fashion.

Chapter 5 : Low defect density WS$_2$ monolayers grown by Chemical Vapor Transport Reaction

5.1 Introduction

The scaling trend of silicon-based electronics faces increasing challenges at sub-10 nm nodes, resulting from the arrival of silicon's physical limitation[10]. Thus, two-dimensional (2D) layered semiconductors, capable of being atomic-thin level without degrading their properties, are proposed as promising candidates for future channel materials. Electrostatic government of 2D channel is easier than the conventional bulk body, which quenches short-channel effects encountered by ultrascaled FETs[127].

Even though competitive 2D FETs with scaled channel dimensions have been demonstrated, the majority of the promising results are based on mechanically exfoliated TMDCs monolayer flakes because it is widely accepted that the electrical quality of mechanically exfoliated sample is superior to those from synthetic processes[77]. However, the non-scalability hinders their practical applications. Therefore, to make TMDCs material move towards real industry, one key challenge is high defect densities presenting in synthetic TMDCs layers from scalable CVD approaches[83, 128, 129]. Although some restoring therapies are reported to fix or passivate the defects[87], it is open to question whether these post-deposition methods can sustain their functions after the sequential device fabrication process, which usually involves thermal and high vacuum steps. Hence, the ultimate solution is suppressing the pristine defect existing in as-grown TMDCs layers. Innovative endeavors on growth reaction to enhance their quality are urgently needed.

Taking WS$_2$ as an example, we report that the chemical vapor transport approach provides an extremely pure metal precursor for more efficient sulfurization in CVD, leading to as-grown WS$_2$ monolayers with a low charge defect. The structural defects of the resulting CVT-WS$_2$ monolayers (~2×10^{12} cm^{-2}) do not increase after typical film transfer processes. By contrast, the WS$_2$ by conventional CVD is vulnerable to the transfer, where the defect density increases from ~8×10^{12} cm^{-2} to ~2×10^{13} cm^{-2}. The bottom-gated field-effect transistor (FET) devices based on the proposed growth reach average electron mobility 100 cm^2/Vs with a peak at ~200 cm^2/Vs at room temperature, comparable to those from exfoliated flakes. The optimized FET device displays a high on-state current ~400 µA/ µm, encouraging the industrialization of 2D materials.

5.2 Experimental Section

5.2.1 Materials synthesis and transfer

CVD-WS$_2$ monolayer samples were synthesized on sapphire substrates by typical CVD method with tungsten oxide (WO$_3$, Sigma-Aldrich, 99.995%) powders and sulfur (S, Sigma-Aldrich, 99.99%) powders as precursors. Generally, the S powders, WO$_3$ powders and sapphire substrates were placed on the upper stream, center and downstream of the furnace. After vacuum the growth chamber under 1mtorr, Ar/H$_2$ gas was purged into the chamber and keep the chamber pressure at 10 torr. The temperature was elevated to 900°C and kept for 15 min for growth. After the growth, the furnace temperature was cooling down naturally.

CVT-WS$_2$ monolayer samples were achieved in a chemical vapor transport system. WS$_2$ powders and sulfur powder were used as precursors. The S powders were placed upstream of the furnace, and an additional heating belt controlled the temperature. The WS$_2$ powder was placed on the center of the furnace while the sapphire substrates were placed downstream. The growth was kept for 15 min and followed by a natural cooling to room temperature. For the wafer-scale growth, the growth pressure was set at 500 torrs in a 3-inch furnace.

The as-grown monolayer WS_2 samples were transferred onto the target substrate for different purposes by a polydimethylsiloxane (PDMS)-assisted transfer method. Generally, the PDMS film was smoothly placed on the top of the WS_2/sapphire sample and soak it into 1M KOH for 5 min. Next, slowly peeled off the PDMS/WS_2 film and transferred it to a target substrate. Anneal the sample for 20 min at 70°C to remove the residue water and increase the adhesion. The last step was slowly peeled off the PDMS film to get a clean WS_2 sample on the target substrate.

5.2.2 Device fabrication and electrical measurements.

For two probe device measurements, we used standard e-beam lithography to pattern the electrodes of WS_2 backgated FETs on Al_2O_3/Si substrates. 20 nm Ti/30 nm Pd was electron beam evaporated as the source/drain/voltage probe contact metal, followed by lift-off. And then, all devices were annealed at 250°C in an ultrahigh vacuum to improve contacts. Electrical measurements were performed by an Agilent B1500 semiconductor parameter analyzer in a close-cycle cryogenic probe station with base pressure $\sim 10^{-5}$ Torr after 350K in situ vacuum annealing at base pressure to remove interface adsorbates.

For the short channel device, the monolayer WS2 films were transferred onto the commercial SiN_x film (thickness = 100 nm) on p^{++}-Si substrates as back gated field-effect transistors (FET). Then Helium-ion beam lithography (ORION NanoFab, Zeiss) with the ion-beam-resist, PMMA (Allresist, AR-P 672-Serie, spin-coated with

4000 rpm for 40s and baked at 180 °C for 3 min.) was used to pattern the source/drain (S/D) metal contacts, which defined the channel length (L_{CH}) from 100 nm to 400 nm and was developed by using 1:3 mixture of 4-methyl-2-pentanone (MIBK) and isopropyl alcohol (IPA). For the contact metals, 20 nm of Bi followed by 15 nm of gold (Au) encapsulating layer were deposited on the WS_2 using e-gun evaporation at a high vacuum chamber ($\sim 1 \times 10^{-7}$ torr). The metal lift-off process was carried out in warm acetone (60 °C) and then rinsed by IPA. Finally, the WS_2 electrical characteristics were measured in a vacuum system (10^{-5} - 10^{-6} Torr) in a Lakeshore probe station using a Keithley 4200-SCS parameter analyzer.

For the four-terminal device measurement, the heavily doped Si substrate was used as a back gate and the 300 nm SiO_2 was used as a gating dielectric. The devices were patterned using PMMA masks and electron beam lithography. 5nm Al and 65nm Au electrodes were deposited using e-beam evaporation. The electrical characterization of monolayer WS_2 FETs were carried out under vacuum ($<10^{-4}$ Torr) in a JANIS CCS350 closed-cycle refrigerator (10-500 K). Our four-terminal measurements were performed from 15 to 300 K, starting from the lowest temperature. The gate and drain biases are provided by the Keithley Model K-6430 Sub-Femtoamp Remote SourceMeters, which are also used to monitor the leakage current and drain current. And the Keithley Model K-2182 is used to sense the voltage difference as a voltmeter. In our structure, the voltage sensed probes minimally affect the current flow in the channel material and thus act like perfect voltmeters.

5.2.3 STM measurements

Our STM experiments were conducted at 77 K in the commercial ultra-high vacuum LT-STM system (CreaTec) with a base pressure of 1.0×10^{-10} mBar. All STM images were acquired in the constant-current mode by using a chemically etched tungsten tip and the bias voltages refer to the sample with respect to the STM tip. Before measurement, the samples were annealed at ~ 550K for over 3 hours to remove possible adsorbates. Note that such a low annealing temperature is used to avoid the transition to sulfur vacancies.

5.3 Results and Discussion

5.3.1 Chemical vapor transport approach

For high-performance electronics in advanced technology nodes, the thickness of transistor channels needs to be as thin as possible to ensure sufficient gate control with dimension scaling. With a thickness around 1 nm, the TMDCs monolayer has been considered a promising channel material. Among the typical TMDCs monolayer, WS_2 provides the best performance for both electrons and holes with high mobilities and saturation velocities[130]. Conventional chemical vapor deposition (CVD) methods can provide scalable WS_2 monolayers through the direct sulfidation of tungsten trioxide (WO_3) or other oxygen-containing precursors. Although single crystal WS_2 flakes can be achieved, pronounced defects are still present, which leads to inferior electrical performance in devices.

We discover that the chemical vapor transport (CVT) approach enables the growth of WS_2 monolayers with significantly lower defect densities. In contrast to the direct sulfidation of WO_3, the CVT method utilizes the transport agent to transport high-purity W source to reduce the oxygen incorporation in deposited WS_2 films, where the setup is schematically depicted in **Figure 5.1a**. The typical optical images of the CVT-grown WS_2 monolayers were shown in **Figure 5.1b**. Their domain size can reach several hundred microns, and the cm-scale continuous WS_2 monolayer film is achievable with an appropriate precursor supply and a suitable growth temperature (**Figure 5.1c**).

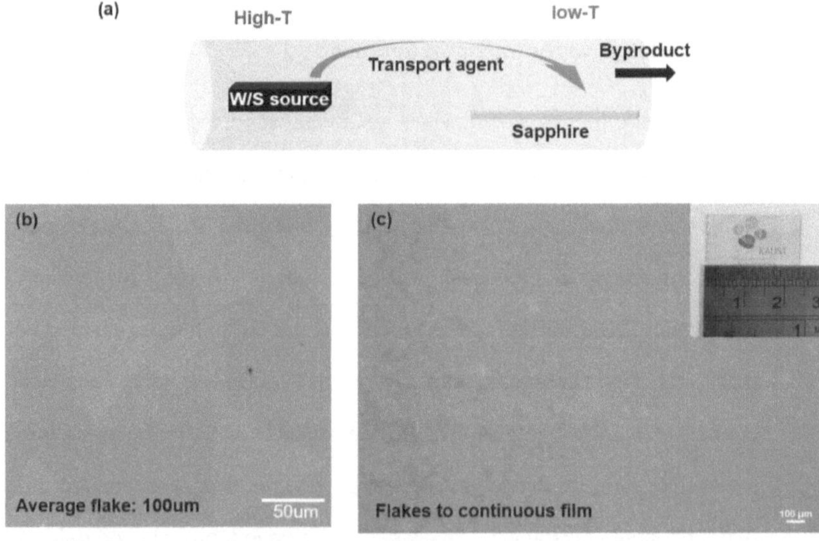

Figure 5.1 Chemical vapor transport deposition of WS_2 monolayer. (a) Schematic of CVT epitaxy growth of WS_2 monolayers. (b) Optical Image image of typical CVT-WS_2 monolayer. (c) Photos of centimeter-scale CVT-WS_2 monolayer film on the sapphire substrate.

5.3.2 High optical quality of CVT-WS$_2$ monolayers

Raman spectra of the as-grown WS$_2$ monolayers prepared by conventional sulfidation of WO$_3$ (labeled as CVD) and the proposed MACVD are compared in **Figure 5.2a**, where many characteristic modes are identified. The prominent peak at ~350 contains in-plane vibration mode E_{2g}^{1} (Γ) (~354 cm^{-1}) and second-order longitudinal acoustic mode 2LA(M). Two defect-sensitive modes including out-of-plane mode A$_{1g}$ and longitudinal acoustic mode LA(M) located at ~416 cm^{-1} and 173 cm^{-1}, respectively[131].

Photoluminescence (PL) spectra of monolayer WS$_2$ is composed of multiple peaks at room temperature, for example, neutral exciton (X^0), trion (X^T) and defect-bound exciton (X^D)[132]. As shown in **Figure 5.2b**, CVT WS$_2$ typically exhibits higher peak energy and narrower full width half maximum (FWHM). The statistical summary of PL spectra for at least 50 single crystal flakes from different growth batches for each case is shown in **Fig. 5.2c.** Conspicuously, the average peak energy of CVT-WS$_2$ monolayers was ~2 eV, which was higher than ~1.98 eV for CVD-WS$_2$ monolayers. The Y-direction shows that the FWHM of CVD-WS$_2$ monolayers was ~65meV, which was larger than that of ~40 meV for CVT-WS$_2$ monolayers. Since X^0 has the highest energy, the PL emission will red-shit with an increase of the FWHM when the materials have a high concentration of dopping X^T and defect X^D emission, which again corroborates the better quality of CVT-WS$_2$[133]. **Figure 5.2d** shows the PL measurement for both samples at 4K deconvoluted by Gaussian functions, where the high energy mode is assigned to X^0, the peak with lower energy by ~30 meV is assigned to X^T, and the broad peak with the lowest energy is assigned to X^D. The significantly lower intensity of X^T and X^D P peaks for CVT-WS2 indicate a lower defect density in it, consistent with the room temperature results.

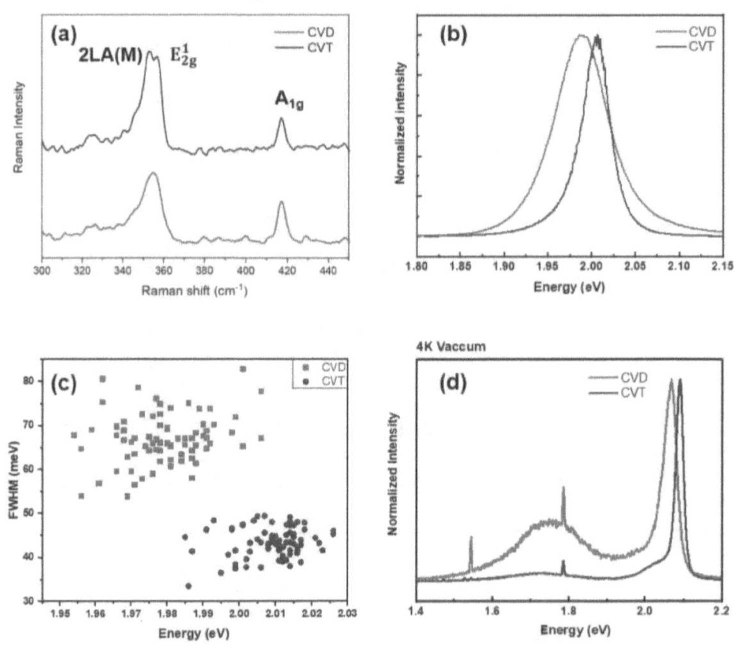

Figure 5.2 Optical characterization of CVT- and CVD-WS$_2$ monolayer. (a) Typical Raman spectra showing the characteristic modes of CVT- and CVD- WS$_2$ monolayers excited by 532 nm wavelengths. (b) Typical PL spectra of CVT- and CVD- WS$_2$ monolayers taken at room temperature. (e) Statistic results of PL peak energy and FWHM for CVT- and CVD-WS$_2$ monolayers. (f) Low-temperature PL spectra of CVT- and CVD- WS$_2$ monolayers at 4K.

5.3.3 Defect density characterization by STM

Following Schuler et al.[134], scanning tunneling microscopy (STM) is adopted to characterize the defect type and defect density for WS_2 monolayers. **Figures 5.3a and 5.3b** show the typical STM images of CVD-WS_2 and CVT-WS_2 monolayers directly grown on conducting highly oriented pyrolytic graphite (HOPG) substrates. It is conspicuous that CVD-WS_2 exhibits a larger amount of point defects than CVT-WS_2. These typically observed point defects can be further categorized into few types, including sulfur substituted by oxygen (O_S), tungsten substituted by molybdenum (Mo_W), and negatively charged defect (CD). Relative to O_S and Mo_W, only a negligible amount of charged defects are found in both samples. It is noted that the density of sulfur vacancies is extremely low and vacancies are almost passivated by oxygen atoms, where we show the identified chance to observe. The detailed scanning tunneling spectroscopy (STS) analysis (**Figure 5.4**) for the major O_S defects, including O_S(top) and O_S(bottom), concludes that these defects do not introduce in-gap states and the STS profiles are close to that in pristine WS_2 regions, agreeing well with previous reports.[85, 134, 135]

To estimate the number density for various defects, more than 20 STM images (40nm by 40nm) for each CVD-WS_2 and CVT-WS_2 are analyzed and **Figure 5.3c** presents the total counts and estimated density of different point defects. The O_S defect is the predominant point defect in both CVD-WS_2 and CVT-WS_2. The estimated O_S defect density of CVD-WS_2 is close to the previous report ($\sim 10^{13}$ cm^{-2})[136], while it was called sulfur vacancy which now has been characterized as oxygen substituting sulfur by STS. The lower Os density in CVT-WS$_2$ indicates a higher degree of sulfidation during the growth. A significantly smaller amount of charged defect (CD), in the order of 10^9 cm^{-2} and 10^{10} cm^{-2}, is found in CVT-WS2 and CVD-WS2. In addition to O$_s$ and CD, we also observe Mo$_W$ defects with the density in the order of 10^{12} cm^{-2} for only CVD-WS$_2$, which is likely caused by the presence of Mo impurity in the W-precursors (~6.5ppm in WO$_3$ according to the material provider). In brief, the as-grown CVT-WS$_2$ exhibits a lower overall defect density, consistent with the Raman and PL measurements.

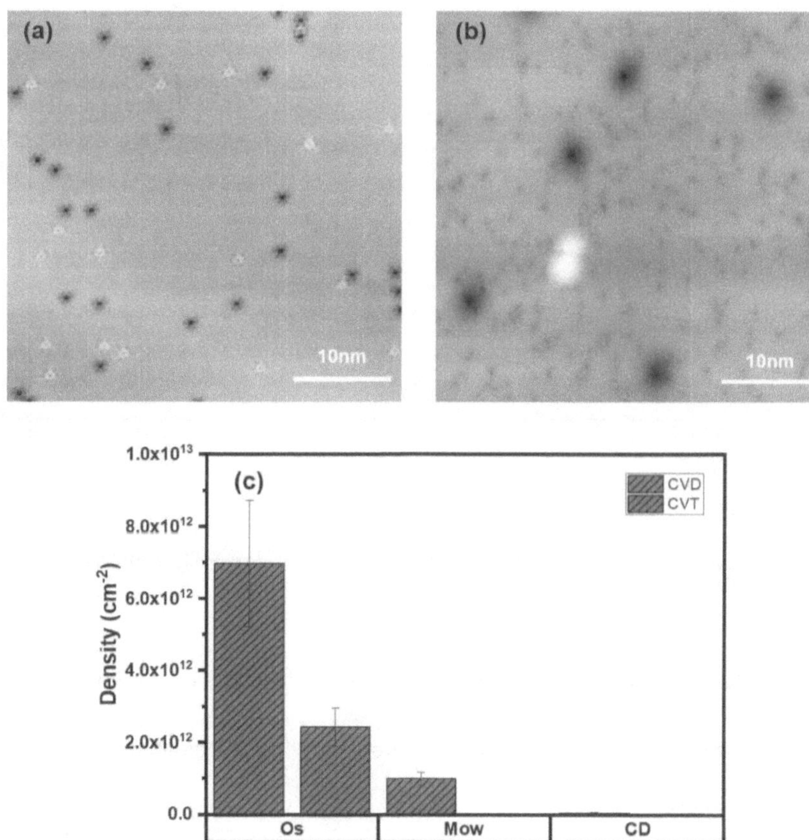

Figure 5.3 Defect analysis by STM technique. STM images of (a) CVD-WS2 and (b) CVT-WS2 monolayer. (c) Histograms of observed point defects in CVT- and CVD- WS$_2$.

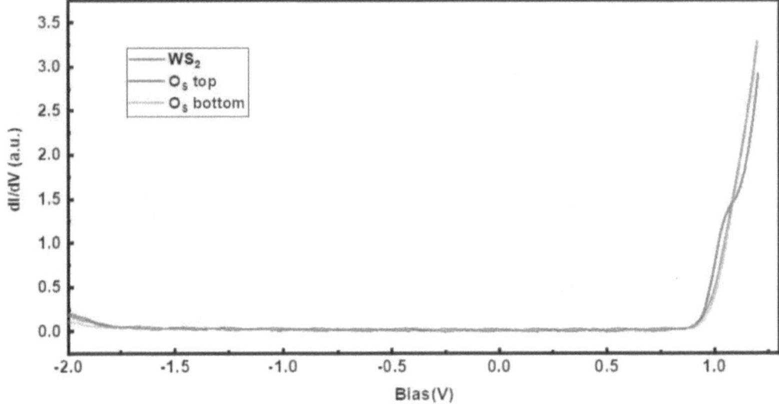

Figure 5.4dI/dV spectrum on the O$_S$ top (red line), O$_S$ bottom (orange line) defect and prinstine WS$_2$ monolayer (gray line).

5.3.4 The transfer process impact on materials quality

Since the WS$_2$ films are typically grown on high-temperature durable sapphire substrates, the transfer of WS$_2$ onto Si-based substrates is needed for device fabrication. Thus, the defect analysis after transfer should be acquired to better correlate to their electrical properties. **Figure 5.5** summarizes the estimated number density of various defects for both samples after transfer processes, corroborating that the CVT-WS$_2$ is better in quality. The density of all types of defects increases after transfer, suggesting that the transfer processes deteriorate the quality of the flakes and delicate transfer approaches should be developed in the future. Recent DFT simulation argues that O$_S$ does not introduce in-gap charged states and only marginally affects WS_2 electronic structures[134]. Also, the band structure of WS_2 with a Mo_W closely resembles that of pristine WS_2.[134] Hence, we suspect that the electron mobility of WS_2 may not be critically affected by the O_S and Mo_W defects. It is noteworthy that the charged defects in CVD-WS_2 increases to $(2.39\pm1.68) \times10^{11}$ cm^{-2} after transfer, significantly larger than the 4.2×10^{10} cm^{-2} in CVT-WS_2.

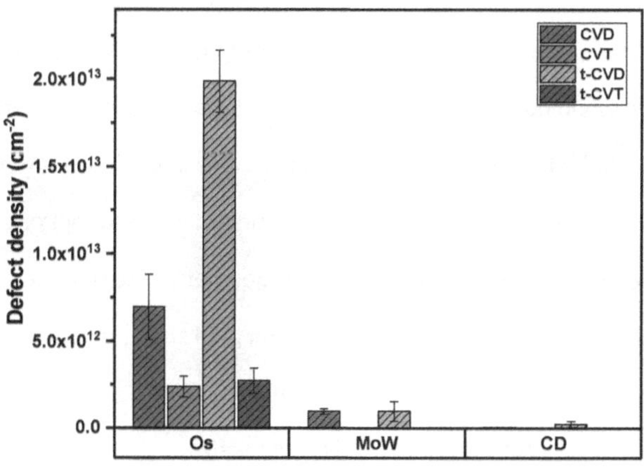

Figure 5.5 STM statistic defect density of CVT- and CVD-WS₂ monolayer before and after transfer.

5.3.5 Electrical properties of CVT-WS₂ monolayers

To evaluate the electrical performance of the CVT-WS_2 monolayers, we fabricated the field-effect transistors with the commonly used back-gate configuration. To begin with, electrical measurements in high vacuum (~ 10^{-6} Torr) with a standard four-probe technique (inset in Figure 4a). **Figure 5.6a** presents the four-probe conductivity $\sigma = (I_d/\Delta V) \times (L_{CH}/W_{CH})$ as a function of backgate voltage V_g at various temperature (200 K to 300 K), where I_{ds} is the source-drain current; ΔV, L_{CH} and W_{CH} are the voltage difference, length and width between the two voltage probes, respectively. The CVT-WS$_2$ sample shows good conductivity at different temperatures. We also extracted the field-effect mobility μ_{FE} from our four-probe measurements, using the expression $\mu_{FE} = (d\sigma/dV_g) \times (1/C_{ox})$, where $C_{ox} = 1.9 \times 10^{-7}$ F cm^{-2}, the geometric gate capacitance per unit area for 30 nm AlO$_x$ dielectric. The corresponding values of temperature-dependent mobility are presented in **Figure 5.6b**. To the extent of our knowledge, this is the highest room-temperature field-effect mobility ($\mu_{FE} = 104$ cm^{2}V^{-1}s^{-1}) reported so far for bottom-up synthesized monolayer WS$_2$ regardless of the device geometry. Furthermore, we summarize the electrical performance of our results and published monolayer WS$_2$ transistors in **Figure 5.6c** for better comparison. The MACVD-WS$_2$ samples display superior performance to conventional CVD synthesized WS$_2$, even comparable to exfoliated samples. To gain further insights into the electrical transport results, we observed that the transfer curves of CVT-WS$_2$ intersect near $V_g = 5.2$ V, which is a trademark of metal-insulator transition (MIT). In general, MIT behavior is more facilely recognized in the low charge-trap-state sample, i.e., as-exfoliated or vacancy-repaired samples.[87] The trap density of CVT-WS$_2$ is about 5.7×10^{12} cm^{-2} ($n_{tr} \approx V_{MIT}C_{ox}/e$), which is quite close to the reported results with specialized interface modification.[137] In addition, we demonstrated short channel FET performance reaching a high on-state current. The inset of **Figure 5.6d** shows the SEM image of 100 nm L_{CH} device. As the measured transfer (I_d-V_g) curves displayed in **Figure 5.6d**, FETs based on CVT-WS$_2$ monolayer can reach a maximum I_{on} ≈ 403 µA/µm and I_{on}/I_{off} current ratio ~ 10^8 at V_{ds} = 1V. The output characteristics at various gate voltages as specified in **Figure 5.6e**, where I_{on} gets to 520 µA/µm at V_{ds} = 1.5V. Based on the above results, it is apparent that the proposed CVT approach pushes the electrical performance of bottom-up synthesized monolayer WS$_2$ to another level, even equivalent to exfoliated samples.

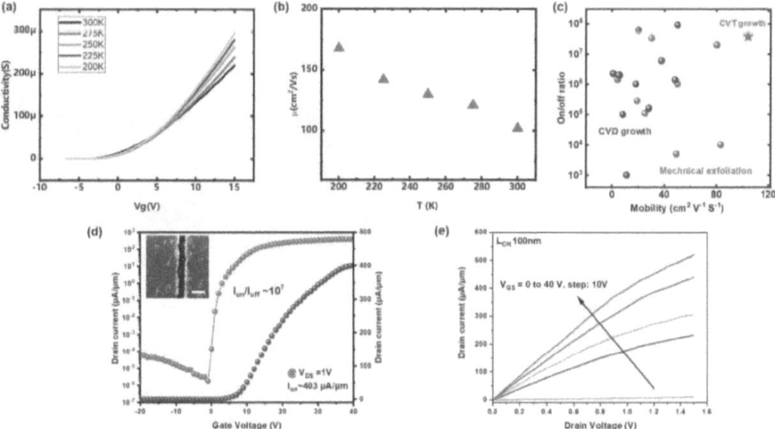

Figure 5.6 Electrical performance of CVT- WS$_2$ monolayer. (a) Four-probe conductivity as a function of Vg for CVT-WS$_2$ monolayer device on the AlO$_x$ substrate at different temperatures. (b) Field-effect mobility as a function of temperature for CVT-WS$_2$ monolayer. (c) Comparison of mobility distribution for our CVT-WS$_2$ results (red) and mechanical exfoliation WS$_2$ monolayers (ME, green), conventional CVD-WS$_2$ (blue) from literatures. (d, e) FET transfer curve and output characteristics of a CVT-WS$_2$ monolayer for the short channel device (L_{CH} = 100nm)

Table 5.1 Device summary from literature.

Growth method	Electrode	Dielectric	Mobility [$cm^2V^{-1}s^{-1}$]	ON/OFF ratio	Ref.	Type
ME	Cr/Au	SiO_2	0.23	-	[138]	$\mu_{2\text{-point}}$
ME	Cr/Au	SiO_2	4	1.43×10^6	[139]	$\mu_{2\text{-point}}$
ME	Cr/Au	SiO_2	5.8	2×10^6	[140]	$\mu_{2\text{-point}}$
KI-doped ME	Cr/Au	SiO_2	30	3.3×10^7	[140]	$\mu_{2\text{-point}}$
ME	Au	Ionic liquid/ SiO_2	19	2.8×10^5	[141]	$\mu_{2\text{-point}}$
ME	Au	SiO_2	50	10^6	[142]	$\mu_{4\text{-point}}$
MPS-ME	Ti/Pd	Al_2O_3/SiO_2	83	10^4	[143]	$\mu_{4\text{-point}}$
ME	Ti/Pd	SiO_2	25	1.1×10^5	[143]	$\mu_{4\text{-point}}$
ME	Ti/Pd	Al_2O_3/SiO_2	49	5×10^3	[143]	$\mu_{4\text{-point}}$
ME	Au/Al	SiO_2	80	2×10^7	[144]	$\mu_{2\text{-point}}$
ME	Au/Al	$WS_2/BN/SiO_2$	163	8.3×10^6	[144]	$\mu_{2\text{-point}}$
ME	Au/Al	$BN/WS_2/BN/SiO_2$	214	3.3×10^6	[144]	$\mu_{2\text{-point}}$
CVD	Al	SiO_2	0.91	2.25×10^6	[145]	$\mu_{2\text{-point}}$
CVD	Ti/Au	SiO_2	5	2×10^6	[146]	$\mu_{2\text{-point}}$
CVD	Ti/Au	SiO_2	8.1	10^5	[147]	$\mu_{2\text{-point}}$
CVD	MLG	SiO_2	50	9×10^7	[146]	$\mu_{2\text{-point}}$
CVD	Ti/Au	HfO_2	48	1.4×10^6	[148]	$\mu_{4\text{-point}}$
CVD	Ti/Au	SiO_2	37.6	6×10^6	[149]	$\mu_{2\text{-point}}$
CVD	Ti/Au	SiO_2	28	1.6×10^5	[133]	$\mu_{2\text{-point}}$
CVD	Ti/Au	SiO_2	20	6×10^7	[150]	$\mu_{2\text{-point}}$
MOCVD	Ti/Au	SiO_2	18	10^6	[16]	$\mu_{4\text{-point}}$
CVD	Ti/Au	SiO_2	11	10^3	[151]	$\mu_{2\text{-point}}$

5.4 Conclusion

To summarize, we have reported a chemical vapor transport route to achieve the ultrahigh grade WS_2 monolayers. From both phonon and photon characteristic aspects, CVT-WS_2 is nearly flawless. Also, the atom-scale inspection of STM and STEM reveals extremely low defect density of as-grown specimens, on a par with fresh-prepared bulk surfaces. The electrical characteristics of the CVT-WS_2 show high average electron mobility with a peak at ~200 cm^2/Vs at room temperature, and over 500 µA/µm on-state current in short channel devices which has never been approached by conventional bottom-up synthetic methods. It is worth noting that the typical MIT behavior generally observed in exfoliated samples emerged as well. Although we exclusively used WS_2 in this work, our method should be compatible with other 2D TMDCs that share similar reaction mechanism. The proposed method paces a further step towards the root issue of 2D electronics development, i.e., reliability and scalability of channel materials. Together with the future progress of unpolluted transferring and fabrication processes that minimize surface impurities, the realization of electronics and low-standby-power integrated circuits based on 2D materials can be foreseen.

Chapter 6 : Conclusions and Perspective

6.1 Conclusions

This book proposed modified CVD approaches to study the growth mechanism and synthesis controllability of 2D TMDCs materials. According to the presented results, we come up with several conclusions.

First, we found that the precursor ratio can tune the tensile strain level of TMDCs materials on sapphire substrates due to the different nucleation formations. As a result, 0.4% ~1.2% tensile trained WSe_2 monolayers were obtained through varying the precursor supply.

Second, a two-step CVD approach was applied to investigate the growth mode transition on the 1st epilayer with different strain levels. We found that high tensile strained 1st epilayer TMDCs directed the layer-by-layer growth mode, critical for the homogeneous epitaxial growth. Consequently, bilayer and trilayer homostructure and heterostructure were successfully achieved. This mechanism is significant for the future application of TMDCs multilayer growth.

Last, a chemical vapor transport (CVT) system was designed for high-quality monolayer TMDCs growth. We found that the transport agent significantly increases the transport ability, which upgrades the quality of epitaxial TMDCs materials. The optical characterization, STM results, and electrical performance demonstrated that our CVT approach provides large-scale high quality and low defect density monolayer WS_2.

6.2 Perspective

6.2.1 Wafer-scale layer controllability of 2D TMDCs

In Chapter 4, we proposed the mechanism of strain-directed layer-by-layer growth; however, the method we used to induce the tensile train could only provide TMDCs flakes with micrometer-size. Therefore, an approach to synthesize large-area high strain level TMDCs film is considerable. A feasible way is utilizing the thermal expansion difference between substrate and TMDCs materials incorporated with the fast cooling process. After we found the possible solution to synthesize large-area high strain TMDCs, a wafer-scale system with the continuous supplement of precursors needs to be designed for multilayer growth, which could be the MOCVD system MACVD system we mentioned in Chapter 5.

6.2.2 Twist-stacked homo- and hetero-structures growth

Recently, the twist angle stacked 2D TMDCs materials have become popular due to the discovery of magic-angle in twisted bilayer graphene[152]. The easiest way to fabricate the twist-angle materials is exfoliation and transferring the materials together. However, a productive method is essential to realize future applications. In chapter 4, we found that the strain-directed bilayer TMDCs preferred 2H- and 3R- stacking sequences which are the most stable configuration. Therefore, the external environment like strain or dopants and sufficient thermal energy could be critical to breaking the rules. More different types and levels of strain should be simulated or experienced to understand deeper mechanisms.

6.2.3 Defect repair technique for 2D TMDCs

In chapter 5, we systematically analyzed the defect types in TMDCs before and after transfer. The oxygen substituted sulfur defects are the dominant defect type in epitaxial grown TMDCs. Therefore how to repair this type of defect become essential for the further improvement of TMDCs' quality.

APPENDICES

Appendix A: Chapter 4 Supplementary Information

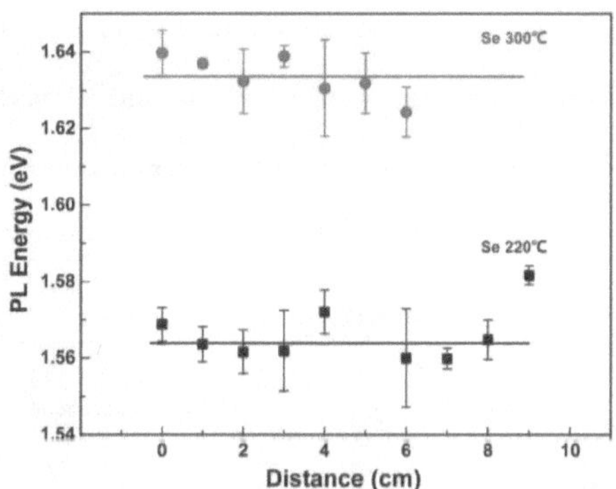

Figure A1 PL energy distribution as a function of 1st WSe$_2$ epilayer epitaxially grown at different distance with respect to the WO$_3$ precursor source. Here we keep the annealing temperature constant near the center of finance while adjusting the Se heating temperature at 220°C (black) and 300°C (red), respectively. Error bars indicate standard deviation of PL energy, measured over five samples at each distance.

Supplementary Discussion 1

As shown in **Figure A2**, the energy of the central peak C is lower than neutral exciton E$_X$ by ~40-50 meV, which is a bit higher than the typical trion binding energy of ~25-30 meV[112]. Therefore, it is not possible to conclude the origin of such a deviation whether stemmed from the trion or defect emission. Note that peak C has a negligible effect on the overall analysis due to the weaker intensity. On the other hand, we carefully checked the temperature dependence in **Figure**

A3. Although PL spectra exhibit low-energy defect emissions, we observed that PL intensities are much weaker than the main E_X peaks for both 1st HTS- and 1st LTS-WSe2 epilayers. Meanwhile, the PL intensity of defect emission decreases rapidly and becomes unobservable when T > 90K. We thus conclude that the main PL peaks originated from the band-to-band transitions, in which the tensile strain causes the energy shift between 1st HTS- and 1st LTS-WSe2 epilayers. This conclusion is consistent with the room-temperature measurement.

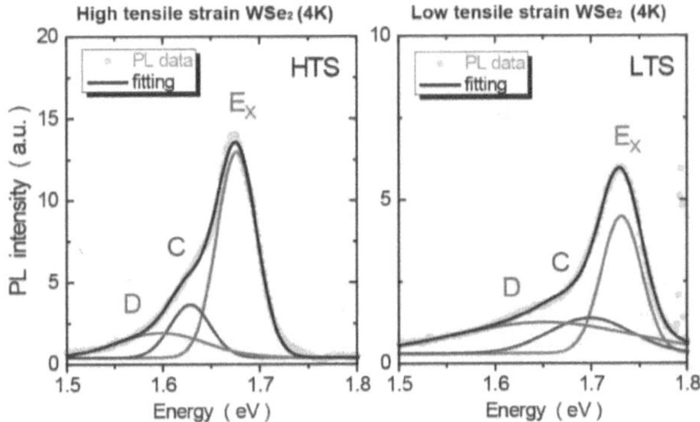

Figure A2 Low temperature PL spectra of 1st HTS- and 1st LTS-WSe2 epilayers. These spectra were fitted with exciton peak (E_X), central peak (C) and defect peak (D).

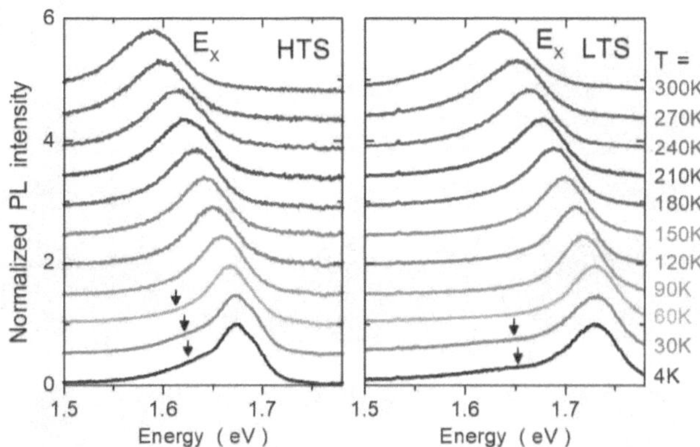

Figure A3. Temperature dependent PL spectra of 1st HTS- and 1st LTS-WSe$_2$ epilayers.

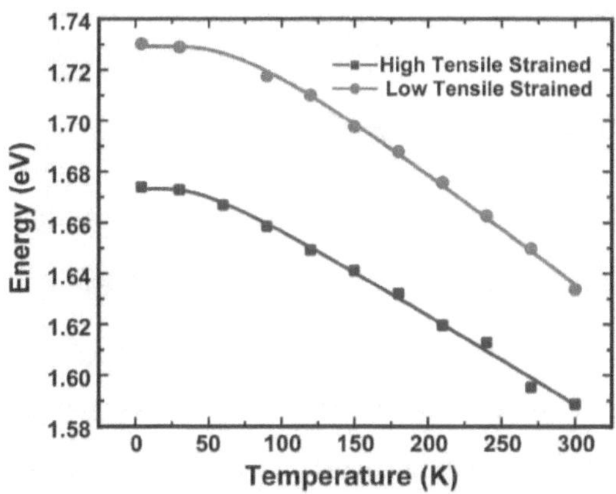

Figure A4. Temperature dependence of the photon energy of both 1st HTS- (blue) and 1st LTS-WSe$_2$ (red) specimens. Data points were fitted based on Eq. (1)

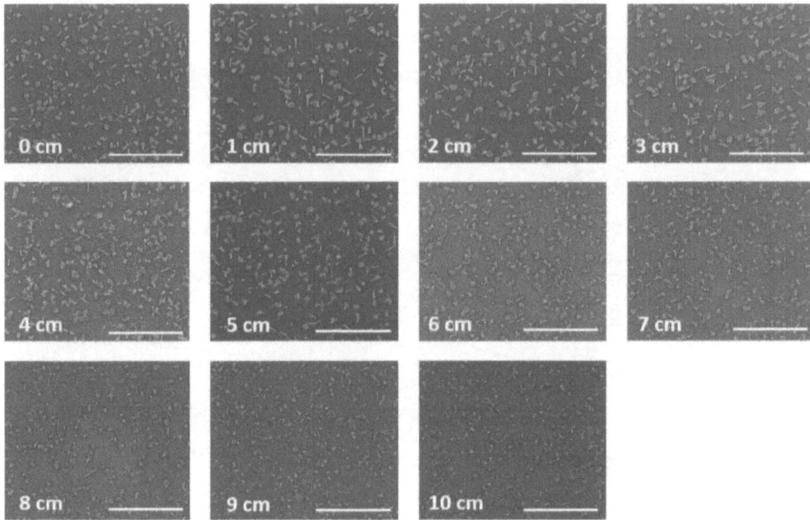

Figure A5. Scanning electron microscope (SEM) images reveal the distance dependent nucleation/deposition of W_xO_y precursors with respect to the WO_3 precursor boat. (Scale bar: 10µm)

Supplementary Discussion 2

The modified growth process was shown in **Figure A6** (top panel). Salt-coated SiO_2/Si substrate was placed on top of the quartz boat in a face-down manner. 0.1g of WO_3 powdery precursors were placed near the downstream end of the quartz boat. A mixture of Ar/H_2 was used as the carrier gas to diffuse the Se vapour towards quartz boat. The furnace temperature was set at 900°C, and the growth pressure was kept constant at 10 torr. As the growth proceeds toward completion, WSe_2 monolayer flakes were found to spatially distribute between the upstream and downstream areas of the SiO_2/Si substrate. PL spectra from 11 different flakes were again collected from both areas as shown in **Figure A6** (bottom panel). Statistical analyses indicate that WSe$_2$ flakes collected in downstream area exhibit a relatively lower PL energy/ higher built-in strain (1%~1.11%) than those sampled from the upstream area (0.77~0.95%). Note that, similar to the epitaxy on sapphire, we again observed the notable concentration gradient of WO$_3$ to Se in the downstream area of SiO$_2$/Si substrate.

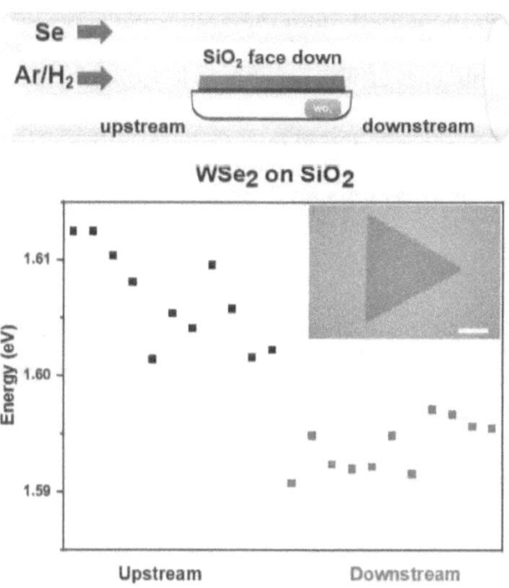

Figure A6. Schematic representation illustrates the epitaxy setup of WSe$_2$ grown on SiO$_2$/Si substrate (top panel). The PL energy distribution of monolayer WSe$_2$ collected from both upstream and downstream areas (11 different flakes at each area) clearly shows the location dependence of built-in strain fields (bottom panel). Inset is the optical microscope image of monolayer WSe$_2$ grown on SiO$_2$/Si substrate. Scale bar: 20 μm.

Supplementary Discussion 3

Mismatch in thermal expansion coefficient (TEC) between the epilayer and growth substrate could also account for the presence of built-in strain. According to the previous reports[71], the theoretical upper limit for the strain from temperature difference can be calculated using the temperature dependent TEC for the WSe_2 (α_{WSe_2}) and substrate (α_{Sub}):

$$\varepsilon(T_g) = \int_{25°C}^{T_g} \alpha_{WSe_2}(T)dT - \int_{25°C}^{T_g} \alpha_{Sub}(T)dT$$

To this end, we measured the temperature variation along the diffusion pathway of our CVD furnace at 900°C as shown in Supporting Information **Table A1** and have arrived the conclusion that built-in strains on 1st WSe_2 epilayers driven by the temperature difference between 0 cm and 9 cm would be less than 0.03%.

Table A1. The temperature varies along the diffusion pathway of our CVD furnace. The position labeled with 0 cm corresponds to the center of the furnace where WO_3 source is located. Here, the growth temperature is set at 900°C.

Position(cm)	0	1	2	3	4	5
T (°C)	927	928	928	927	925	921
Position(cm)	8	9	10	11	12	13
T (°C)	901	892	882	868	852	828

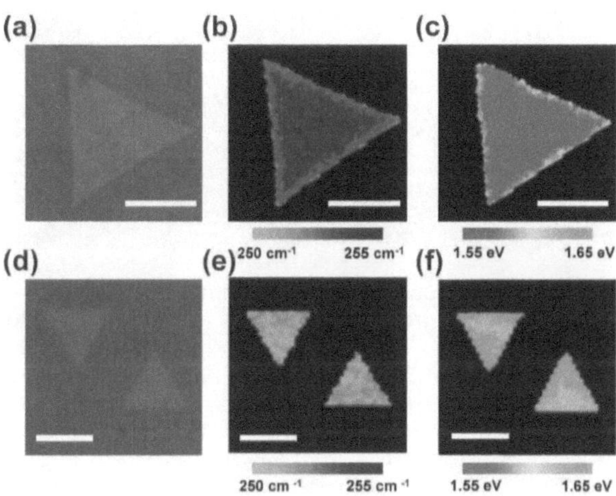

Figure A7 (a, d) Optical microscope image, (b, e) Raman mapping and (c,f) PL mapping of 1st HTS- (top panel) and 1st LTS-WSe$_2$ epilayers (bottom panel). Scale bar: 5 µm.

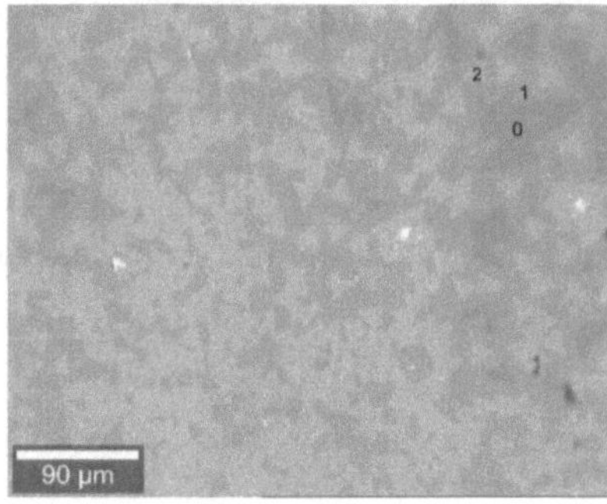

Figure A8. Optical microscope image of WSe$_2$ homobilayers. The label of 0 denotes the Al$_2$O$_3$ substrate, 1 stands for monolayer and 2 represents bilayer.

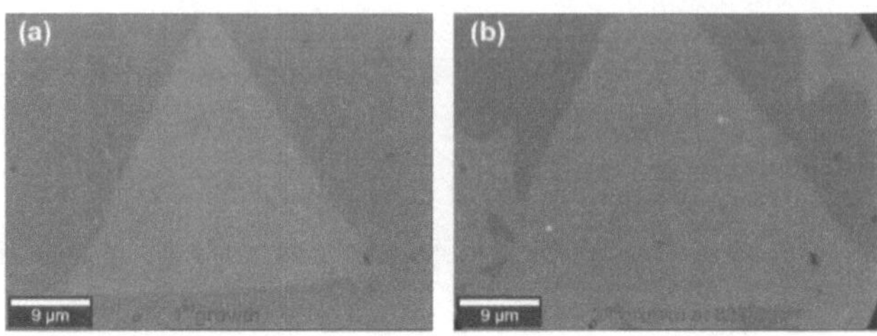

Figure A9. Optical microscope images of (a) 1st HTS-WSe$_2$ epilayer and (b) 2nd WSe$_2$ overlayer. The growth temperature (surface temperature of substrates) is set at 825°C.

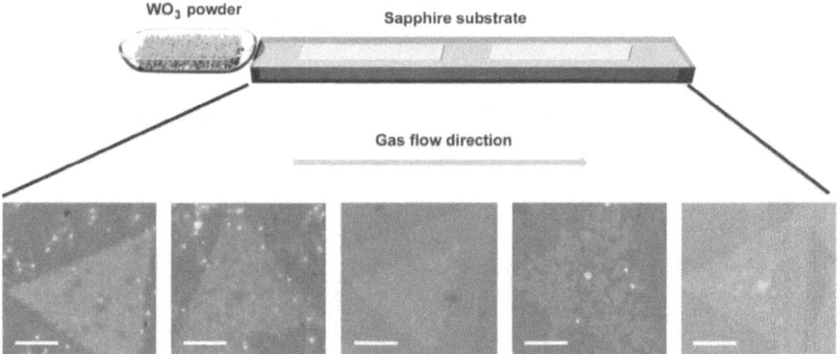

Figure A10. Optical microscope images reveal the growth variation of 2nd WSe$_2$ overlayers collected from upstream (left) to downstream (right). Scale bar: 10 μm.

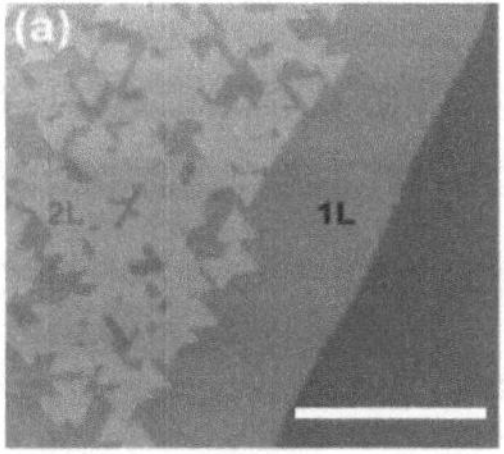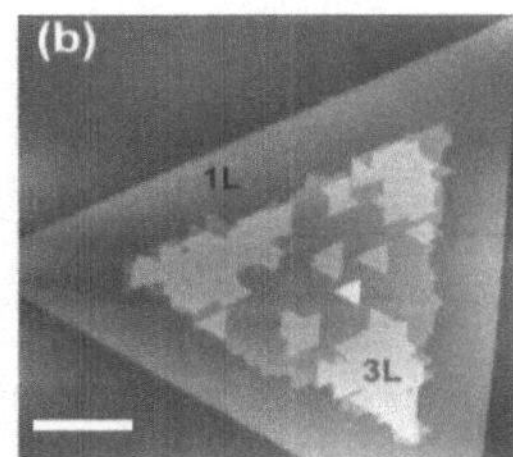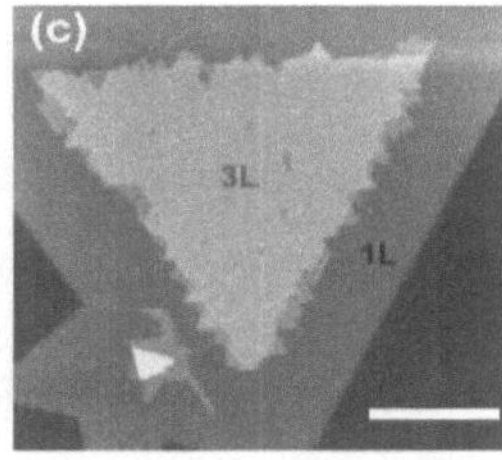

Figure A11. AFM images of (a) triangular WSe$_2$ (2L) flakes grown on monolayer (1L) WSe$_2$, (b) triangular WSe$_2$ flakes (3L) grown on WSe$_2$ homo- bilayer and (c) fully covered trilayer. No island-like clusters were observed. Scale bar: 5 μm.

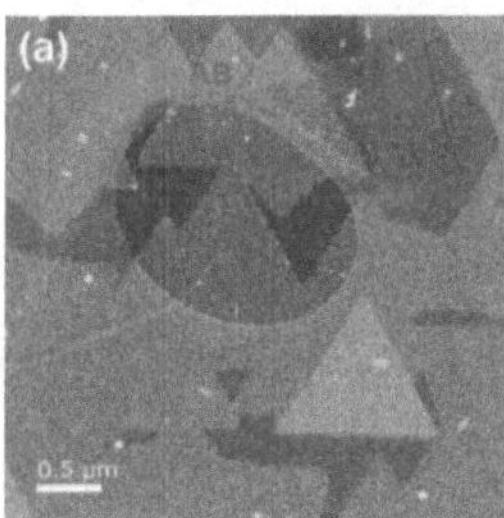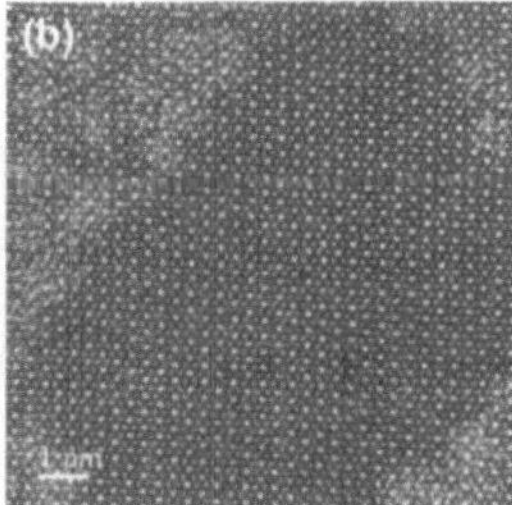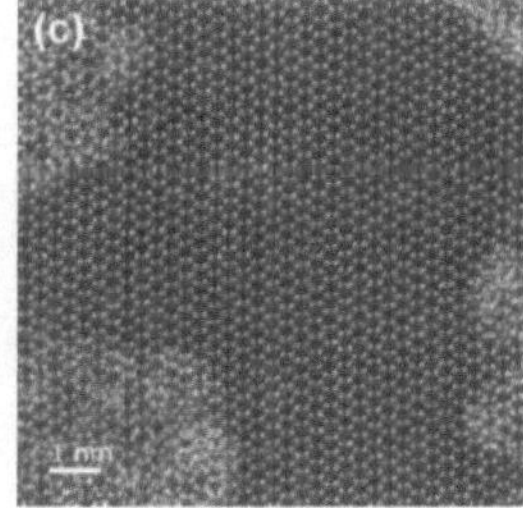

Figure A12. (a) ADF-STEM image of as grown WSe$_2$ bilayer triangles. Low resolution HAADF-STEM image of (b) AB stacking and (c) AA' stacking WSe$_2$ bilayer.

Supplementary Discussion 4

The optical properties which are strongly associated with the quality of the WSe_2 homo-bilayers are investigated by PL, and Raman spectra systematically. **Figure A12a** shows the typical PL spectra taken from monolayer (1L), bilayer (2L) and trilayer (3L) WSe_2 flakes. Noted that the 1L here refers to the lateral growth from the edge of the 1st HTS-WSe_2 epilayer during the formation 2nd WSe_2 overlayer. Since the highly strained 1st HTS-WSe_2 epilayer becomes relaxed during the inception of epitaxy of 2nd WSe_2 overlayer, the PL peak A emanated from the newly grown 1L is characterized by the emergence of a direct bandgap of 1.63eV. Meanwhile, we observed a transition to an indirect bandgap on 2L WSe_2 homostructures as is evident in the emission peak I emerging at around 1.54eV along with a broad shoulder of 1.6eV. The intensity of peak I attenuates considerably with a minor red-shift to 1.52eV when the number of WSe_2 overlayers increases to 3L. This transition in peak I can be attributed to a unique layer dependence of bandgap evolution in WSe_2. Figure S12b shows the Raman spectra of 1L WSe_2 along with its multilayered counterparts at regions near the low-frequency (left) and the predominant E_{2g}/A_{1g} peaks (right). Two dominant peaks are found to emerge around 250 cm^{-1} and do not show a layer-dependence. Meanwhile, another peak located at 308 cm^{-1} is assigned to B_{2g} active mode which originates from the presence of interlayer interaction in 2L and 3L WSe_2. In the low-frequency sector of Raman spectra, the shear mode (S) at 16 cm^{-1} and the layer-breathing mode (LB) at 28 cm^{-1} appears when examining the 2L WSe_2, and both peaks notably shift to and combine at ~20 cm^{-1} upon the formation of the 3L

WSe$_2$. Here, the finding of directing the growth front through activating the basal plane during the epitaxy has made possible the preservation of all these characteristics unique to the exfoliated benchmarks, thus distinguishing the strain-engineering approach from the other strategies.[153]

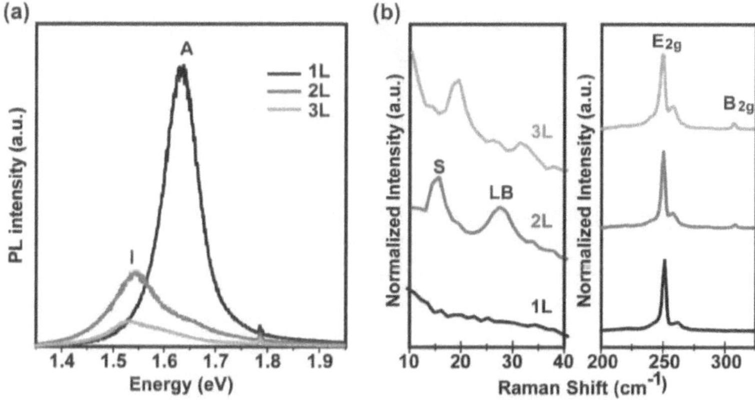

Figure A13. (a) Photoluminescence spectra of the as-grown monolayer (1L), bilayer (2L), and trilayer (3L) WSe$_2$. (b) Raman spectra of 1L, 2L, and 3L-WSe$_2$ in the ultra-low frequency range and in the range of the E_{2g}^1.

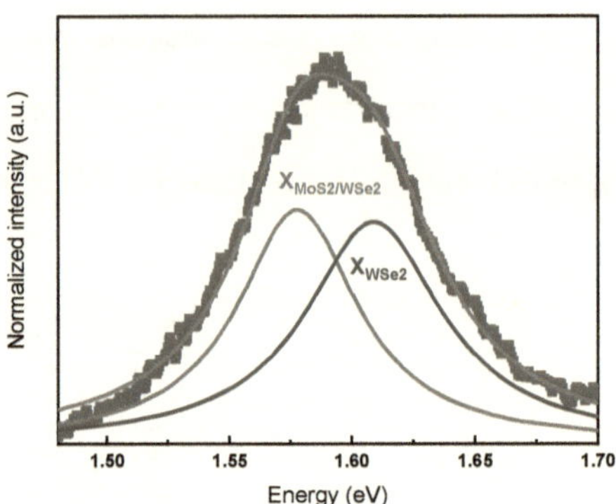

Figure A14. Deconvolution of PL emission from MoS_2/WSe_2 hetero-bilayer reveals the exitonic emission originated from WSe_2 (curve in blue) and interlayer exciton between MoS_2/WSe_2 (curve in green), respectively.

www.ingramcontent.com/pod-product-compliance
Lightning Source LLC
LaVergne TN
LVHW041710190726
843493LV00007B/2026